Teach Yourself
VISUALLY™

Calculus

Visual

by Dale Johnson

WILEY

Wiley Publishing, Inc.

Teach Yourself VISUALLY™ Calculus

For general information on our other products and services or to obtain technical support please contact our Customer Care Department within the U.S. at (800) 762-2974, outside the U.S. at (317) 572-3993 or fax (317) 572-4002.

Wiley also publishes its books in a variety of electronic formats. Some content that appears in print may not be available in electronic books. For more information about Wiley products, please visit our web site at www.wiley.com.

Library of Congress Control Number: 2008924083

ISBN: 978-0-470-18560-5

Printed in the United States of America

10 9 8 7 6 5 4 3 2 1

Book production by Wiley Publishing, Inc. Composition Services

Praise for the Teach Yourself VISUALLY Series

I just had to let you and your company know how great I think your books are. I just purchased my third Visual book (my first two are dog-eared now!) and, once again, your product has surpassed my expectations. The expertise, thought, and effort that go into each book are obvious, and I sincerely appreciate your efforts. Keep up the wonderful work!

—Tracey Moore (Memphis, TN)

I have several books from the Visual series and have always found them to be valuable resources.

—Stephen P. Miller (Ballston Spa, NY)

Thank you for the wonderful books you produce. It wasn't until I was an adult that I discovered how I learn—visually. Although a few publishers out there claim to present the material visually, nothing compares to Visual books. I love the simple layout. Everything is easy to follow. And I understand the material! You really know the way I think and learn. Thanks so much!

—Stacey Han (Avondale, AZ)

Like a lot of other people, I understand things best when I see them visually. Your books really make learning easy and life more fun.

—John T. Frey (Cadillac, MI)

I am an avid fan of your Visual books. If I need to learn anything, I just buy one of your books and learn the topic in no time. Wonders! I have even trained my friends to give me Visual books as gifts.

—Illona Bergstrom (Aventura, FL)

I write to extend my thanks and appreciation for your books. They are clear, easy to follow, and straight to the point. Keep up the good work! I bought several of your books and they are just right! No regrets! I will always buy your books because they are the best.

—Seward Kollie (Dakar, Senegal)

Credits

Acquisitions Editor
Pam Mourouzis

Project Editor
Suzanne Snyder

Copy Editor
Kelly Henthorne

Technical Editor
Tom Page

Editorial Manager
Christina Stambaugh

Publisher
Cindy Kitchel

Vice President and Executive Publisher
Kathy Nebenhaus

Interior Design
Kathie Rickard
Elizabeth Brooks

Cover Design
José Almaguer

Dedication

I would like to dedicate this book to the memory of Dr. Jerry Bobrow, who passed away on November 12, 2007. Jerry was a long time author of math preparation books published by CliffsNotes™/Wiley. I had the privilege of working for and with Jerry since 1980. He introduced me to the excitement of math preparation courses, both teaching them and assisting in writing books to be used in these courses. It is because of Jerry's careful mentoring of me and his undying encouragement, that this book has become a reality for me. His smile, laughter, enthusiasm, and energy will remain a part of my life for ever. I'd like to offer special thanks to his wife Susan, daughter Jennifer, and sons Adam and Jonathan for sharing Jerry with me and so many others—all of whose lives have been enriched by having known Jerry Bobrow.

About the Author

Dale Johnson taught middle school and high school mathematics for 35 years, retiring in 2005. He received a BA in Mathematics in 1971 from the University of California, Riverside and an MA in Education in 1979 from San Diego State University.

Since 1980, he has taught the Math Review portion of test preparation courses for the CBEST, CSET, GRE, and GMAT through California State University Extended Studies at San Diego, Fullerton, Long Beach, and Pomona, as well as at the Claremont colleges.

Dale is the co-author of *CliffsAP Calculus AB and BC*; technical editor for *Calculus For Dummies, Calculus For Dummies Workbook,* and *CliffsQuickReview: Calculus.* He was also a consultant and contributing author for the following CliffsTestPrep titles: *GMAT, ELM & EPT, ACT, SAT, CAHSEE,* and *TAKS,* as well as *Math Review for Standardized Tests.*

Dale resides in Encinitas, California, spending free time with his wife Connie reading, gardening, working out at the gym, and traveling. He goes to many off-road races in California, Nevada, and Mexico with his son Zachary who does computerized designs of the vehicles which race in these events.

Acknowledgments

I want to offer thanks to acquisitions editor Pam Mourouzis for her persistence and encouragement in talking me into taking on this project. Her assistance in mastering the fine points of the use of high tech manuscript preparation was invaluable.

Project editor Suzanne Snyder helped me in the process of shaping the rough manuscript into a form that would be useful to the book's readers. Her many author queries and suggestions for changes led to a more user-friendly format for the book.

The folks in the graphics art department were able to turn my hand-drawn colored pencil illustrations and figures into works of art—thus adding great visual clarity to the printed text. My thanks go out to these amazing graphic artists.

Tom Page, the technical editor, did a masterful job of finding math errors that I made in the rough manuscript. His attention to each detail in illustrations or in solutions to problems was greatly appreciated.

Special thanks to my wife Connie who spent many hours by herself, waiting for me to type "just one more page," and who put up with my writing and/or correcting the manuscript during many moments of our vacation travels.

Table of Contents

An Introduction to Limits

Limits in Calculus 2

Definition of the Limit of a Function 14

One-Sided Limits 17

Determine Limits from the Graph of a Function 20

Calculate Limits Using Properties of Limits 23

Continuity at a Point or on an Interval 26

The Intermediate Value and Extreme Value Theorems 32

chapter 2 *Algebraic Methods to Calculate Limits*

Direct Substitution 36

Indeterminate Forms $\pm\frac{\infty}{\infty}$ and $\frac{0}{0}$ 38

Dealing with Indeterminate Forms 39

Limits at Infinity: Horizontal Asymptotes 48

chapter 3 Introduction to the Derivative

What Can Be Done With a Derivative? 56

Derivative as the Slope of a Tangent Line. 58

Derivative by Definition 60

Find the Equation of a Line Tangent to a Curve 67

Horizontal Tangents 68

Alternate Notations for a Derivative 70

Derivative as a Rate of Change 72

Differentiability and Continuity 74

chapter 4 Derivatives by Rule

Derivatives of Constant, Power, and Constant Multiple 78

Derivatives of Sum, Difference, Polynomial, and Product 80

The General Power Rule 84

The Quotient Rule. 86

Rolle's Theorem and the Mean Value Theorem 89

Limits: Indeterminate Forms and L'Hôpital's Rule 93

chapter 5   *Derivatives of Trigonometric Functions*

Derivatives of Sine, Cosine, and Tangent 97

Derivatives of Secant, Cosecant, and Cotangent. 100

L'Hôpital's Rule and Trigonometric Functions 102

The Chain Rule 104

Trigonometric Derivatives and the Chain Rule. 109

Derivates of the Inverse Trigonometric Functions 110

chapter 6   *Derivatives of Logarithmic and Exponential Functions*

Derivatives of Natural Logarithmic Functions 113

Derivatives of Other Base Logarithmic Functions 119

Logarithms, Limits, and L'Hôpital's Rule 123

Derivatives of Exponential Functions. 125

chapter 7   *Logarithmic and Implicit Differentiation*

Logarithmic Differentiation 130

Techniques of Implicit Differentiation 134

Applications of Implicit Differentiation 139

chapter 8 *Applications of Differentiation*

Tangent Line to Graph of a Function at a Point 143

Horizontal Tangents 144

Critical Numbers. 146

Increasing and Decreasing Functions 148

Extrema of a Function on a Closed Interval 155

Relative Extrema of a Function: First Derivative Test 160

Concavity and Point of Inflection. 165

Extrema of a Function: Second Derivative Test 172

chapter 9 *Additional Applications of Differentiation: Word Problems*

Optimization 177

Related Rates 183

Position, Velocity, and Acceleration 188

Introduction to the Integral

Antiderivatives: Differentiation versus Integration. 195

The Indefinite Integral and Its Properties 197

Common Integral Forms. 201

First Fundamental Theorem of Calculus. 203

The Definite Integral and Area 205

Second Fundamental Theorem of Calculus 209

Techniques of Integration

Power Rule: Simple and General 212

Integrals of Exponential Functions 220

Integrals That Result in a Natural Logarithmic Function 223

Integrals of Trigonometric Functions 226

Integrals That Result in an Inverse Trigonometric Function. 232

Combinations of Functions and Techniques 235

Algebraic Substitution 237

Solving Variables Separable Differential Equations 240

chapter 12 *Applications of Integration*

Acceleration, Velocity, and Position 247

Area between Curves: Using Integration 250

Volume of Solid of Revolution: Disk Method 260

Volume of Solid of Revolution: Washer Method 268

Volume of Solid of Revolution: Shell Method 275

Appendix *283*
Index *287*

chapter 1

An Introduction to Limits

This chapter discusses the importance of limits to the study of both differential and integral calculus. **Differential calculus** involves finding a derivative—such as the slope of a tangent line or the rate of change of a balloon's volume with respect to its radius—of the maximum or minimum value of a function. **Integral calculus** involves finding an integral—such as determining the velocity function from its acceleration function, calculating the area under a curve, finding the volume of an irregular solid, or determining the length of an arc along a curve. Starting with some examples of how you can use limits in calculus, I then introduce an intuitive notion of limits. From the formal definition of a limit, you learn ways to determine limits of functions from their graphs, as well as how to use some basic limit properties. The chapter concludes with a brief discussion of continuity and two important theorems related to continuity.

Limits in Calculus 2

Definition of the Limit of
   a Function 14

One-Sided Limits 17

Determine Limits from the Graph
   of a Function 20

Calculate Limits Using Properties
   of Limits 23

Continuity at a Point or
   on an Interval 26

The Intermediate Value and
   Extreme Value Theorems 32

This section gives you some examples of how to use algebraic techniques to compute limits. These include the terms of an infinite series, the sum of an infinite series, the limit of a function, the slope of a line tangent to the graph of a function, and the area of a region bounded by the graphs of several functions.

TERMS OF AN INFINITE SERIES

1 Let's take a look at the series $1, \frac{1}{2}, \frac{1}{4}, \frac{1}{8}, \frac{1}{16}, \frac{1}{32}, \dots, \frac{1}{2^{n-1}}$ where n is a positive integer. As n gets larger and larger, the term $\frac{1}{2^{n-1}}$ gets smaller and smaller.

$$1, \frac{1}{2}, \frac{1}{4}, \frac{1}{8}, \frac{1}{16}, \frac{1}{32}, \dots, \frac{1}{1024}, \dots, \frac{1}{524,288}, \dots$$
$$\text{for } n = 11 \quad \text{for } n = 20$$

2 If n were large enough (say n approached ∞), it appears that the terms approach 0. In the language of limits, you can say that the *limit* of $\frac{1}{2^{n-1}}$, as n approaches ∞, is 0.

$$\lim_{n \to \infty} \frac{1}{2^{n-1}} = 0$$

LIMIT OF A SUM OF AN INFINITE SERIES

1 Let's go one step further and try to find the sum of the terms of the series mentioned earlier, as n gets very large.

$$1 + \frac{1}{2} + \frac{1}{4} + \frac{1}{8} + \frac{1}{16} + \frac{1}{32} + \dots + \frac{1}{2^{n-1}} + \dots$$

❷ For increasing values of n, the sum of that number of terms is shown at the right.

for $n = 1 \rightarrow sum = 1$

for $n = 2 \rightarrow sum = 1 + \frac{1}{2} = 1\frac{1}{2}$

for $n = 3 \rightarrow sum = 1 + \frac{1}{2} + \frac{1}{4} = 1\frac{3}{4}$

for $n = 4 \rightarrow sum = 1 + \frac{1}{2} + \frac{1}{4} + \frac{1}{8} = 1\frac{7}{8}$

for $n = 8 \rightarrow sum = 1 + \frac{1}{2} + \frac{1}{4} + \frac{1}{8} + \frac{1}{16} + \frac{1}{32} + \frac{1}{64} + \frac{1}{128} = 1\frac{127}{128}$

❸ It appears that the sum of the terms of this series is approaching 2. In the language of limits, we say the limit of the sum of the terms $\frac{1}{2^{n-1}}$, as n approaches ∞, is 2.

TIP

Remember that the symbol Σ (sigma) represents "the sum of."

$$\lim_{t \to \infty}\left(\sum_{n=1}^{t} \frac{1}{2^{n-1}} \right) = 2$$

LIMIT OF A FUNCTION

① The graph of $f(x) = (x + 3)(x - 2)^2$ is shown at the right. It appears that as x gets closer and closer to 2 (from both the left and the right), $f(x)$ gets closer and closer to 0.

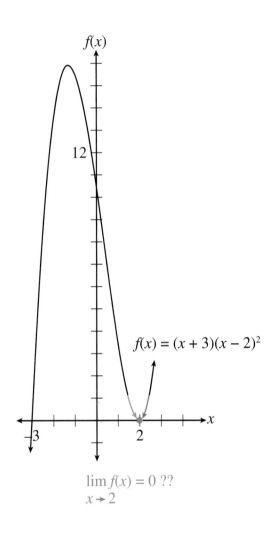

$f(x) = (x + 3)(x - 2)^2$

$$\lim_{x \to 2} f(x) = 0 \ ??$$

❷ Try some values for x close to 2, finding their y coordinates to verify that the limit really is 0.

x and $f(x)$ Values for $f(x) = (x + 3)(x - 2)^2$	
x	$f(x) = (x + 3)(x - 2)^2$
0.5	7.875
1.0	4
1.5	1.125
1.8	0.192 x approaches 2 from the left
1.9	0.049
1.99	0.0005
1.999	0.000005
2	0
2.001	0.000005
2.01	0.0005
2.1	0.051
2.2	0.208 x approaches 2 from the right
2.5	1.375
3.0	6
3.5	14.625

❸ From the chart, it appears that as x gets closer and closer to 2, the value of $f(x)$ gets closer and closer to 0.

$$\lim_{x \to 2} f(x) = 0$$

SLOPE OF LINE TANGENT TO A CURVE

1 The graph of $f(x) = (x - 2)^2 + 1$ is shown at the right with a line tangent to the curve drawn at the point with x-coordinate 3.

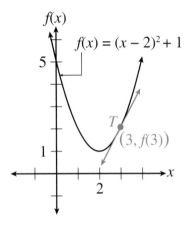

$f(x)$

$f(x) = (x - 2)^2 + 1$

$(3, f(3))$

2 Let's approximate the slope of that red tangent line. Select some values of x that approach 3 from the right side: 4, 3.5, 3.1, 3.01, and, of course, 3. Letting Δx (read "delta x") equal the difference between the selected value of x and 3, you can complete the chart at the right.

	Δx	$3 + \Delta x$	$f(3 + \Delta x)$	Resulting Point
1	4	5	(4,5)	A
0.5	3.5	3.25	(3.5,3.25)	B
0.1	3.1	2.21	(3.1,2.21)	C
0.01	3.01	2.0201	(3.01,2.0201)	D
0	3	2	(3,2)	T

Selected Points of the Graph of $f(x) = (x - 2)^2 + 1$

❸ Next, compute the slopes of the secant lines $\overleftrightarrow{AT}$, $\overleftrightarrow{BT}$, $\overleftrightarrow{CT}$, and $\overleftrightarrow{DT}$.

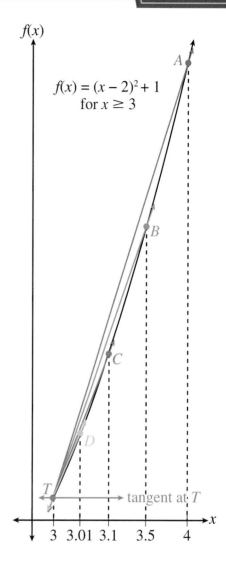

$f(x)$

$f(x) = (x-2)^2 + 1$
for $x \geq 3$

A

B

C

D

T

tangent at T

x

3 3.01 3.1 3.5 4

slope of $\overleftrightarrow{AT} = \dfrac{5-2}{4-3} = 3$

slope of $\overleftrightarrow{BT} = \dfrac{3.25-2}{3.5-3} = \dfrac{1.25}{0.5} = 2.5$

slope of $\overleftrightarrow{CT} = \dfrac{2.21-2}{3.1-3} = \dfrac{0.21}{0.1} = 2.1$

slope of $\overleftrightarrow{DT} = \dfrac{2.0201-2}{3.01-3} = \dfrac{0.0201}{0.01} = 2.01$

4 As the chosen points A, B, C, and D get closer and closer to point T ($\Delta x \to 0$), the slope of the line tangent at $x = 3$, gets closer and closer to 2.

At $x = 3$, the slope of the line tangent to the graph of $(x) = (x–2)^2 + 1$ is 2.

5 For any point P close to T, the slope of $\overleftrightarrow{PT}$

is given by $\quad \dfrac{f(3+\Delta x)-f(3)}{(3+\Delta x)-3}$

$$= \dfrac{f(3+\Delta x)-f(3)}{\Delta x}$$

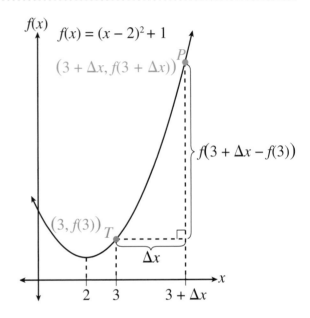

$f(x)$

$f(x) = (x - 2)^2 + 1$

P

$(3 + \Delta x, f(3 + \Delta x))$

$f(3 + \Delta x - f(3))$

$(3, f(3))$ $\quad T$

Δx

x

2 $\quad$ 3 $\quad$ 3 + Δx

6 As $\Delta x \to 0$, the point P moves extremely close to the point T; in this case, the slope of the line tangent at point T will be the expression in Step 5 above.

$$\lim_{\Delta x \to 0} \frac{f(3+\Delta x)-f(3)}{\Delta x} = 2$$

7 The expression in Step 6 in the right column is also known as the *derivative of* $f(x)$ *at* $x = 3$, and is denoted by $f'(3)$. In Chapters 3–6, you will learn many techniques for determining the derivative of a function.

$$f'(3) = \lim_{\Delta x \to 0} \frac{f(3+\Delta x)-f(3)}{\Delta x}$$

Therefore, the slope of the line tangent to the graph of $f(x) = (x - 2)^2 + 1$ at $x = 3$ is 2.

RIEMANN SUM: AREA UNDER A CURVE

❶ The last limit example involves approximating the area below the graph of $f(x) = x^2$, above the x-axis, right of the line $x = 1$, and left of the line $x = 5$.

Note: Try to find both a lower and an upper approximation to the actual area. A lower approximation uses inscribed rectangles and an upper approximation uses circumscribed rectangles.

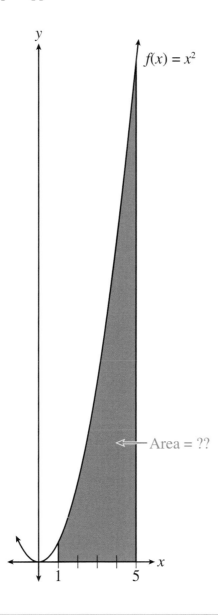

❷ Using four inscribed rectangles, each having a base of 1 unit, their corresponding heights are found: $f(1) = 1$, $f(2) = 4$, $f(3) = 9$, and $f(4) = 16$. The area computation is shown at the right.

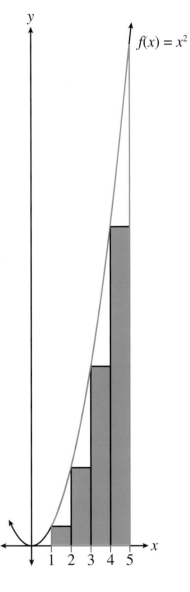

$$\text{Area} = 1 \cdot f(1) + 1 \cdot f(2) + 1 \cdot f(3) + 1 \cdot f(4)$$
$$= 1 \cdot 1 + 1 \cdot 4 + 1 \cdot 9 + 1 \cdot 16$$
$$= 30$$

This area approximation is *less* than the actual desired area.

❸ Next, using four circumscribed rectangles, each having a base of 1 unit, their corresponding heights are found: $f(2) = 4$, $f(3) = 9$, $f(4) = 16$, and $f(5) = 25$. The area computation is shown at the right.

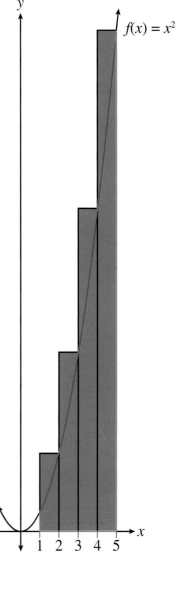

$f(x) = x^2$

$$\begin{aligned} \text{Area} &= 1 \cdot f(2) + 1 \cdot f(3) + 1 \cdot f(4) + 1 \cdot f(5) \\ &= 1 \cdot 4 + 1 \cdot 9 + 1 \cdot 16 + 1 \cdot 25 \\ &= 54 \end{aligned}$$

This area computation is *greater* than the actual desired area.

4 The actual area of the region described in Step 1 is greater than 30 and less than 54. If you wanted a closer approximation of the actual area, you would use a very large number of rectangles, each having a base of $\Delta x = \frac{5-1}{n}$, where n is the number of rectangles used. The corresponding height for each rectangle would then be $f(x_i)$, where i represents the 1st, or 2nd, or 3rd, or 4th rectangle of the n rectangles used.

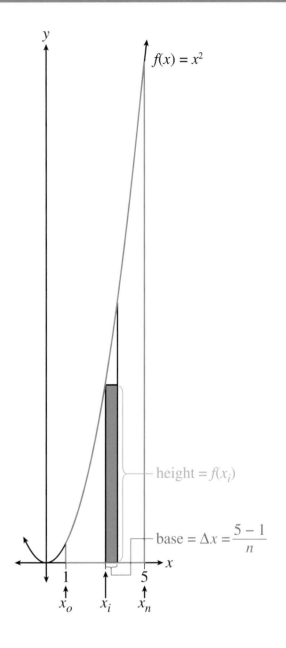

$f(x) = x^2$

height $= f(x_i)$

base $= \Delta x = \dfrac{5-1}{n}$

5 The sum of the areas of these rectangles is represented by the expression at the right, an example of what's called a *Riemann Sum,* the use of which is to predominately find the sum of areas of rectangles under a curve. The area of one green rectangle (see previous page) would be $f(x_i) \cdot \Delta x$.

$$\sum_{i=1}^{n} f(x_i) \cdot \Delta x = \text{sum of the areas of all } n \text{ rectangles.}$$

---

6 The actual area would be found by letting $n \to \infty$, so that $\Delta \to 0$, and then finding the limit of the Riemann Sum.

$$Area = \lim_{n \to 0} \left(\sum_{i=1}^{n} f(x_i) \Delta x \right)$$

---

7 If $f(x)$ is defined on a closed interval [a,b] and $\lim_{n \to \infty} \left(\sum_{i=1}^{n} f(x_i) \Delta x \right)$ exists, the function $f(x)$ is said to be *integrable on [a,b]* and limit is denoted by $\int_a^b f(x) dx$.

$$\lim_{n \to \infty} \left(\sum_{i=1}^{n} f(x_i) \Delta x \right) = \int_a^b f(x) dx$$

---

8 The expression $\int_a^b f(x) dx$ is called the definite integral of f from a to b.

In our example, $Area = \int_1^5 x^2 dx$.

---

9 In Chapter 12, you will compute these sorts of areas, after learning some techniques of integration.

$$Actual\ Area = \int_1^5 x^2 dx = 41\frac{1}{3}$$

Definition of the Limit of a Function

This section introduces the precise definition of the limit of a function and discusses its use in determining or verifying a limit.

THE Δ–E DEFINITION OF THE LIMIT OF A FUNCTION

Let f be a function defined for numbers in some open interval containing c, except possibly at the number c itself. The limit of $f(x)$ as x approaches c is L, written as $\lim_{x \to c} f(x) = L$, if for any $\varepsilon > 0$, there is a corresponding number $\delta > 0$ such that if $< |x - c| < \delta$, then $|f(x) - L| < \varepsilon$.

---

❶ Let's break down the definition of the limit as stated above. Since $|x - c|$ is the distance between x and c, and $|f(x) - L|$ is the distance between $f(x)$ and L, the definition could be worded: $\lim_{x \to c} f(x) = L$, meaning that the distance from $f(x)$ to L can be made as small as we like by making the distance from x to c sufficiently small (but not 0).

As $x \to c$, then $f(x) \to L$, so that $\lim_{x \to c} f(x) = L$.

---

❷ Note that $0 < |x - c| < \delta$ implies that x lies in the interval $(c - \delta, c)$ or in $(c, c + \delta)$. Also, $|f(x) - 9| < \varepsilon$ implies that L lies in the interval $(L - \varepsilon, L + \varepsilon)$.

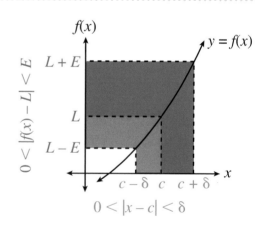

$$0 < |x - c| < \delta$$

Using the δ–ε Definition to Verify a Limit

Use the δ–ε definition to verify that $\lim_{x \to 3} x^2 = 9$.

❶ You must show that for any $\boldsymbol{\varepsilon} > 0$, there corresponds a $\delta > 0$ such that: $|f(x) - 9| < \varepsilon$ whenever $< |x - 3| < \delta$. Since your choice of δ depends on your choice of ε, you need to find a connection between $|x^2 - 9|$ and $|x - 3|$.

$$|x^2 - 9| = |x + 3||x - 3|$$

If you move left and right of $x = 3$ just 1 unit, x would be in the interval (4,5) so that $< |x + 3| < 8$

❷ Let δ be ε / 8.

It follows that when $0 < |x - 3| < \delta = \dfrac{\varepsilon}{8}$

the result is $|x^2 - 9| = |x + 3||x - 3|$

$$< 8\left(\frac{\varepsilon}{8}\right)$$

FINDING A VALUE OF Δ, GIVEN A SPECIFIC VALUE OF E

Given that $\lim_{x \to 2} (3x - 1) = 5$, find a value of δ such that $|(3x - 1) - 5| < 0.01$ whenever $|x - 2| < \delta$.

❶ First, find a connection between $|(3x - 1) - 5|$ and $|x - 2|$.

$$|(3x - 1) - 5| = |3x - 6| = 3|x - 2|$$

❷ You are given that $\left|(3x-1)-5\right|<0.01$.

$$\left|(3x-1)-5\right|<0.01$$
$$3\left|x-2\right|<0.01$$
$$\left|x-2\right|<\frac{.01}{3}$$

❸ Select $\delta=\frac{.01}{3}$.

This choice of δ works since $0<\left|x-2\right|<\delta$ implies that

$$\left|(3x-1)-5\right|=3\left|x-2\right|$$
$$<3(\delta)$$
$$<3\left(\frac{.01}{3}\right)$$
$$<0.01 \text{ the given requirement.}$$

Here, I illustrate limits from the left and from the right—commonly known as one-sided limits.

Notation for One-Sided Limits, with Examples

❶ For the function $f(x) = \frac{1}{x}$ at the right, there is *no* $\lim_{x \to 0} f(x)$. Notice that as $x \to 0$ from the left, $f(x) \to -\infty$ but as $x \to 0$ from the right, $f(x) \to +\infty$.

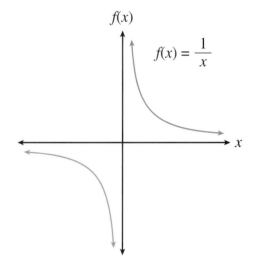

$f(x)$

$f(x) = \dfrac{1}{x}$

x

❷ Each limit in Step 1 is called a **one-sided limit**.
"The limit of $f(x)$ as x approaches c from the left is L" is written as:

$$\lim_{x \to c^-} f(x) = L$$

"The limit of $f(x)$ as x approaches c from the right is M" is written as:

$$\lim_{x \to c^+} f(x) = M$$

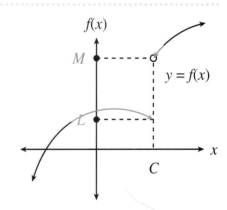

$f(x)$

M

$y = f(x)$

L

C

x

ONE-SIDED LIMITS FOR A RATIONAL FUNCTION

For $f(x) = \dfrac{x^2}{(x-2)^2}$, find $\lim\limits_{x \to 2^-} f(x)$ and $\lim\limits_{x \to 2^+} f(x)$.

From the graph, $\lim\limits_{x \to 2^-} f(x) = +\infty$ and $\lim\limits_{x \to 2^+} f(x) = +\infty$.

Note: *When you write* $\lim\limits_{x \to c} f(x) = \infty$, *it does not mean that the limit exists.*

The limit actually does not exist because f(x) increases without bound as x approaches c.

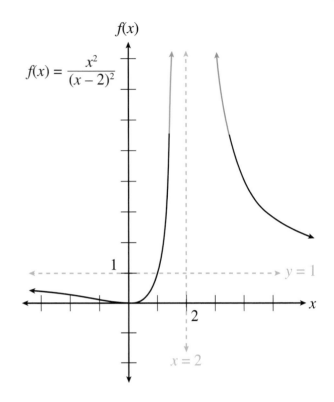

$$f(x) = \dfrac{x^2}{(x-2)^2}$$

$y = 1$

$x = 2$

··

ONE-SIDED LIMITS FOR A CONDITIONAL FUNCTION

For the function $f(x) = x$ if $x \geq 0$, but $f(x) = -x - 1$, if $x < 0$, find $\lim\limits_{x \to 0^-} f(x)$ and $\lim\limits_{x \to 0^+} f(x)$.

❶ From the graph at right, you can see that $\lim\limits_{x \to 0^-} = -1$ and $\lim\limits_{x \to 0^+} f(x) = 0$.

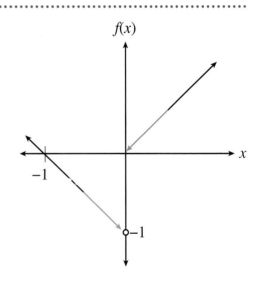

❷ Notice that since the left-sided and right-sided limits are different, $\lim\limits_{x \to 0} f(x)$ does not exist.

ONE-SIDED LIMITS FOR A POLYNOMIAL FUNCTION

For the function $f(x) = \dfrac{x^3}{6} - \dfrac{x^2}{4} - 3x$, find $\lim\limits_{x \to -2^-} f(x)$ and $\lim\limits_{x \to -2^+} f(x)$.

❶ From the graph at the right, you can

see that $\lim\limits_{x \to -2^-} f(x) = 3\dfrac{2}{3}$ and

$\lim\limits_{x \to -2^+} f(x) = 3\dfrac{2}{3}$.

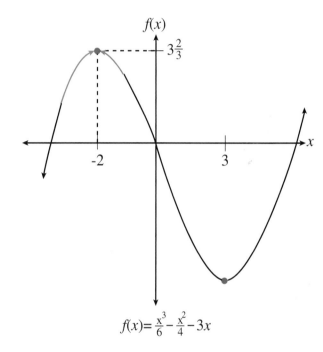

$$f(x) = \frac{x^3}{6} - \frac{x^2}{4} - 3x$$

❷ Since the left-sided and right-sided limits are the same, we can say that $\lim\limits_{x \to -2} f(x) = 3\dfrac{2}{3}$.

Determine Limits from the Graph of a Function

Using just the graph of a function, you can determine one-sided limits, as shown in this section.

LIMITS FROM THE GRAPH OF A FUNCTION: FIRST EXAMPLE

Using the graph of the function $f(x)$ at the right, determine each limit below.

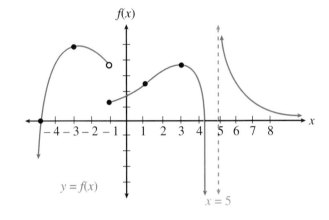

$\lim_{x \to -3} f(x)$

$\lim_{x \to -3} f(x) = 4$

As x approaches -3, $f(x)$ approaches 4.

$\lim_{x \to -1^+} f(x)$

$\lim_{x \to -1^+} f(x) = 1$

As x approaches -1 from the right, $f(x)$ approaches 1.

$\lim_{x \to -1^-} f(x)$

$\lim_{x \to -1^-} f(x) = 3$

As x approaches -1 from the left, $f(x)$ approaches 3.

$\lim\limits_{x \to -1} f(x)$

$\lim\limits_{x \to -1} f(x)$ is nonexistent

As x approaches -1 from the left and then from the right, two different limits are concountered.

$\lim\limits_{x \to 3} f(x)$

$\lim\limits_{x \to 3} f(x) = 3$

As x approaches 3 (from the left or the right), $f(x)$ approaches 3.

$\lim\limits_{x \to 5^-} f(x)$

$\lim\limits_{x \to 5^-} f(x) = -\infty \,(\text{or nonexistent})$

As x approaches 5 from the left, $f(x)$ decreases without limit (approaches $-\infty$).

$\lim\limits_{x \to \infty} f(x)$

$\lim\limits_{x \to \infty} f(x) = 0$

As x gets really large, $f(x)$ appears to be getting really small.

LIMITS OF A TRIGONOMETRIC FUNCTION

To the right is the graph of $f(x) = \sin x$. Determine each limit below.

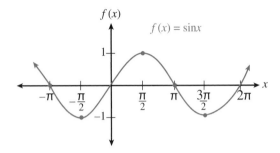

$f(x)$

$f(x) = \sin x$

$\lim\limits_{x \to 0} (\sin x)$

$\lim\limits_{x \to 0} (\sin x) = 0$

$\lim\limits_{x \to \pi^-} (\sin x)$

$\lim\limits_{x \to \pi^-} (\sin x) = 0$

$\lim\limits_{x \to -\pi^+} (\sin x)$

$\lim\limits_{x \to -\pi^+} (\sin x) = 0$

$\lim\limits_{x \to \infty} (\sin x)$

$\lim (\sin x)$ is nonexistent, $\sin x$ oscillates between -1 and 1.

There are many properties that enable you to calculate such limits as sums, products, powers, and even composites of functions.

Properties of Limits

Listed below are the more common limit properties and a corresponding example in the right column. Let k and c be constants, let n be a positive integer, and let f and g be functions such that $\lim\limits_{x \to c} f(x) = L$ and $\lim\limits_{x \to c} = M$.

Let $f(x) = x^2 + 1$ and $g(x) = \frac{1}{x}$

Scalar Product:

$$\lim_{x \to c}\left[k \cdot f(x)\right]$$
$$= k \cdot \left[\lim_{x \to c} f(x)\right]$$
$$= k \cdot L$$

$$\lim_{x \to 2}\left[5 \cdot f(x)\right]$$
$$= 5 \cdot \left[\lim_{x \to 2} f(x)\right]$$
$$= 5(2^2 + 1)$$
$$= 25$$

Sum or Difference:

$$\lim_{x \to c}\left[f(x) \pm g(x)\right]$$
$$= \lim_{x \to c} f(x) \pm \lim_{x \to c} g(x)$$
$$= L \pm M$$

$$\lim_{x \to -1}\left[f(x) - g(x)\right]$$
$$= \lim_{x \to -1} f(x) - \lim_{x \to -1} g(x)$$
$$\left[(-1)^2 + 1\right] - \left(\frac{1}{-1}\right)$$
$$= 2 + 1$$
$$= 3$$

Product:

$$\lim_{x \to c}\left[f(x) \cdot g(x)\right]$$
$$= \lim_{x \to c} f(x) \cdot \lim_{x \to c} g(x)$$

$$\lim_{x \to .5}\left[f(x) \cdot g(x)\right]$$
$$= \lim_{x \to .5} f(x) \cdot \lim_{x \to .5} g(x)$$
$$= \left[(.5)^2 + 1\right] \cdot \left[\frac{1}{.5}\right]$$
$$= (1.25) \cdot 2$$
$$= 2.5$$

Quotient:

$$\lim_{x \to c}\left[\frac{f(x)}{g(x)}\right]$$

$$=\frac{\lim_{x \to c} f(x)}{\lim_{x \to c} g(x)}$$

$$=\frac{L}{M} \text{ if } M \neq 0$$

$$\lim_{x \to 3}\left[\frac{f(x)}{g(x)}\right]$$

$$=\frac{\lim_{x \to 3} f(x)}{\lim_{x \to 3} g(x)}$$

$$=\frac{3^2+1}{\frac{1}{3}}$$

$$=30$$

Power:

$$\lim_{x \to x}\left[f(x)\right]^n$$

$$=\left[\lim_{x \to c} f(x)\right]^n$$

$$\lim_{x \to 2}\left[f(x)\right]^3$$

$$=\left[\lim_{x \to 2} f(x)\right]^3$$

$$=\left(2^2+1\right)^3$$

$$=125$$

Composite:

$$\lim_{x \to c} f\left(g(x)\right)$$

$$=f\left(\lim_{x \to c} g(x)\right)$$

$$=f(M) \text{ if } \lim_{x \to M} f(x) = f(M)$$

$$\lim_{x \to 4} f\left(g(x)\right)$$

$$=f\left(\lim_{x \to 4} g(x)\right)$$

$$=f\left(\frac{1}{4}\right)$$

$$=\left(\frac{1}{4}\right)^2+1$$

$$=1\frac{1}{16}$$

Special Trigonometric Limits:

$$\lim_{x \to 0}\left(\frac{\sin x}{x}\right) = 1$$

$$\lim_{x \to 0}\left(\frac{\sin 3x}{x}\right)$$

$$= \lim_{x \to 0}\left(\frac{\sin 3x}{x}\right) \cdot \frac{3}{3}$$

$$= 3 \cdot \lim_{3x \to 0}\left(\frac{\sin 3x}{3x}\right)$$

$$= 3 \cdot 1$$

$$= 3$$

$$\lim_{x \to 0}\left(\frac{1 - \cos x}{x}\right) = 0$$

$$\lim_{x \to 0}\left(\frac{1 - \cos^2 x}{x}\right)$$

$$= \lim_{x \to 0}\frac{(1 + \cos x)(1 - \cos x)}{x}$$

$$= \lim_{x \to 0}(1 + \cos x) \cdot \left(\frac{1 - \cos x}{x}\right)$$

$$= \lim_{x \to 0}(1 + \cos x) \cdot \lim_{x \to 0}\left(\frac{1 - \cos x}{x}\right)$$

$$= 1 \cdot 0$$

$$= 0$$

Continuity at a Point or on an Interval

This section discusses continuity of a function along with methods for determining continuity. It also introduces some applications of continuity, including the Intermediate Value Theorem and the Extreme Value Theorem.

Definition of a Function Continuous at a Point

❶ The function f is continuous at the number c if the following conditions are satisfied:

i) $f(c)$ exists

ii) $\lim\limits_{x \to c} f(x)$ exists

iii) $\lim\limits_{x \to c} f(x) = f(c)$

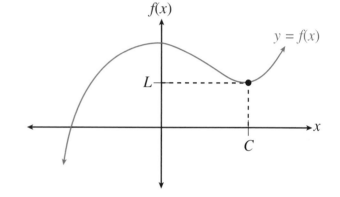

i) $f(c) = L$

ii) $\lim\limits_{x \to c} f(x) = L$

iii) $\lim\limits_{x \to c} f(x) = L = f(x)$

❷ A practical test for continuity means you can sketch the graph of the function without lifting your pencil off the paper. Another test for continuity by viewing the function's graph is that the graph has no *holes*, no *jumps,* and no *vertical asymptotes*.

An example of the first figure is

$$f(x) = \frac{x^4 - 5x^3 + 9x^2 - 5x - 2}{x - 2}$$

$$= \frac{(x - 2)(x^3 - 3x^2 + 3x + 1)}{x - 2}$$

An example of the second figure is

$$f(x) = \begin{cases} (x - 1)^2 + 2 \text{ if } x > 2 \\ -(x - 1)^2 - 3 \text{ if } x \leq 2 \end{cases}$$

An example of the third figure is

$$f(x) = \frac{1}{(x - 1)^3} + 1$$

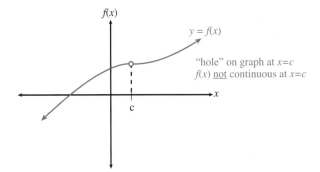

$f(x)$

$y = f(x)$

"hole" on graph at $x=c$
$f(x)$ <u>not</u> continuous at $x=c$

c

x

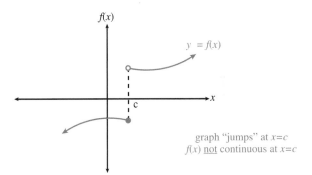

$f(x)$

$y = f(x)$

c

x

graph "jumps" at $x=c$
$f(x)$ <u>not</u> continuous at $x=c$

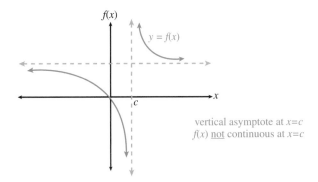

$f(x)$

$y = f(x)$

c

x

vertical asymptote at $x=c$
$f(x)$ <u>not</u> continuous at $x=c$

REMOVABLE DISCONTINUITY

Given a function f which is not continuous at some number: the discontinuity at c is called *removable* if f can be made continuous at c by defining or redefining $f(c)$ so that the function f is continuous *at c*.

The graph of $f(x) = \begin{cases} x \text{ if } x \neq 3 \\ 5 \text{ if } x = 3 \end{cases}$ is shown at the

right. Explain why $f(x)$ is not continuous at $x = 3$.

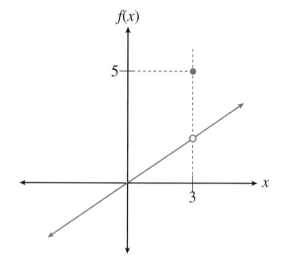

This is done by checking each of the three conditions needed for a function to be continuous at $x = 3$.

i) $f(3)$ exists

ii) $\lim\limits_{x \to 3} f(x)$ exists

iii) $\lim\limits_{x \to 3} f(x) = f(3)$

i) $f(3) = 5$

ii) $\lim\limits_{x \to 3} f(x) = 3$

iii) $\lim\limits_{x \to 3} f(x) = 3 \neq 5 = f(3)$

Condition iii of the definition fails. Note the "hole" at $x = 3$. This is also known as a **removable discontinuity.**

JUMP DISCONTINUITY

The graph of $f(x) = \begin{cases} x \text{ if } x \geq 2 \\ -1 \text{ if } x < 2 \end{cases}$ is shown at the right.

Explain why $f(x)$ is not continuous at $x = 2$.

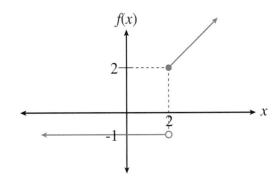

i) $f(2) = 2$

ii) $\lim\limits_{x \to 2} f(x)$ is nonexistent

Condition ii) of the definition fails. Note the "jump" at $x = 2$. This is also known as a **jump discontinuity.**

INFINITE DISCONTINUITY

The function f is said to have an *infinite discontinuity at c if*
$$\lim\limits_{x \to c} f(x) = \infty \text{ or } \lim\limits_{x \to c} f(x) = -\infty.$$

Note: *x can approach c from the left or from the right.*

The graph of $f(x) = \dfrac{x}{x-1}$ is shown at right.

Explain why $f(x)$ is not continuous at $x = 1$.

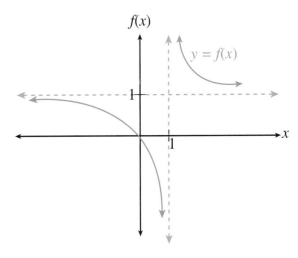

i) $f(1)$ is not defined

ii) $\lim\limits_{x \to 1} f(x)$

Conditions i) and ii) of the definition fail. Note the "vertical asymptote" at $x = 1$. This is also known as an **infinite discontinuity.**

Properties of Continuity

If k is a real number and functions $f(x)$ and $g(x)$ are continuous at $x = c$, then the following functions are also continuous at $x = c$.

Scalar Multiple: $k \cdot f(x)$ and $k \cdot g(x)$ **Quotient:** $\dfrac{f(x)}{g(x)}$, provided that $g(c) \neq 0$

Sum and Difference: $f(x) \pm g(x)$

Product: $f(x) \cdot g(x)$ **Composite:** $f\big(g(x)\big)$ if $f(x)$ is continuous at $g(c)$

ADDITIONAL CONTINUOUS FUNCTIONS

❶ Polynomial functions are continuous everywhere:

$$p(x) = a_n x^n + a_{n-1} x^{n-1} + \; a_2 x^2 + a_1 x^1 + a_0$$

$f(x) = 5x^7 - 13x^3 + 4x - 5$ is continuous everywhere.

❷ The following functions are continuous at every point in their domain.

Rational: $r(x) = \dfrac{f(x)}{g(x)}$ where $g(x) \neq 0$

$f(x) = \dfrac{x^2 + 5x + 17}{x - 1}$ is continuous everywhere except at $x = 1$.

Radical: $f(x) = \sqrt[n]{x}$

$f(x) = \sqrt{x + 1}$ is continuous for $x \geq -1$.

Trigonometric: $\sin x$, $\cos x$, $\tan x$, $\csc x$, $\sec x$, $\tan x$

$f(x) = \tan x = \dfrac{\sin x}{\cos x}$ is continuous except at $x = \dfrac{\pi}{2} + k\pi$ where k is an integer.

Exponential: $f(x) = n^x$

$f(x) = 3^{x-2}$ is continuous everywhere.

Logarithmic: $f(x) = \log x$ or $f(x) = \ln x$:

$f(x) = \log(1 - x)$ is continuous when $x < 1$.

DEFINITION OF CONTINUITY ON AN OPEN INTERVAL

A *function* is continuous on an open interval (a,b) if and only if it is continuous at every number c in (a,b) (i.e., no holes, jumps, or vertical asymptotes in (a,b)).

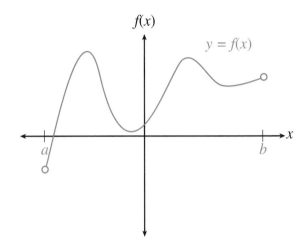

$f(x)$ is continuous on the open interval (a,b).

DEFINITION OF CONTINUITY ON A CLOSED INTERVAL

A function $f(x)$ *is continuous on the closed interval [a,b]* if it is continuous on the open interval (ab) and $\lim\limits_{x \to a^+} f(x) = f(a)$ and

$$\lim_{x \to b^-} f(x) = f(b).$$

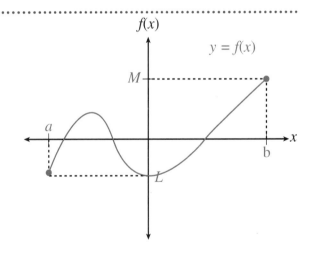

$f(x)$ is continuous on the open interval (a,b)

i) $\lim\limits_{x \to a^+} f(x) = L$ and $f(a) = L$

ii) $\lim\limits_{x \to b^-} f(x) = M$ and $f(b) = M$

Therefore, $f(x)$ is continuous on the closed interval $[a,b]$.

The Intermediate Value and Extreme Value Theorems

There are two theorems that you will find useful: The Intermediate Value Theorem and the Extreme Value Theorem.

The Intermediate Value Theorem is this: If a function f is continuous on $[a,b]$ and m is any number between $f(a)$ and $f(b)$, then there is at least one number, c, between a and b, for which $f(c) = m$.

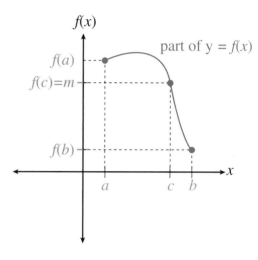

Using the Intermediate Value Theorem to Prove the Existence of a Zero of a Function in a Given Interval

Show that the function $f(x) = x^3 - x^2 - 2x$ has at least one zero in the interval $[1,2]$ and then find that zero.

❶ Find $f(1)$ and $f(2)$.

$$f(x) = x^3 - x^2 - 2x$$
$$f(1) = 1^3 - 1^2 - 2(1) = -2$$
$$f(3) = 3^3 - 3^2 - 2(3) = 12$$

2 Apply the conclusion of the Intermediate Value Theorem.

Since $f(1) < 0$ and $f(3) > 0$, there must be at least one number c in $[-2, 12]$ for which $f(c)=0$.

3 Set the original function equal to 0.

$0 = x^3 - x^2 - 2x$

4 Factor the right-hand side.

$0 = x(x^2 - x - 2)$
$0 = x(x - 2)(x + 1)$

5 Solve for x.

$x = 0, x = 2, x = -1$

The only value of x the interval $(1,3)$ is 2. Therefore, $c = 2$.

The Intermediate Value and
Extreme Value Theorems *(continued)*

Extreme Value Theorem

If a function *f* is continuous on [*a,b*], then *f* has both a maximum value and a minimum value on [*a,b*].

Case I: *f* has both a minimum and a maximum value on (*a,b*).

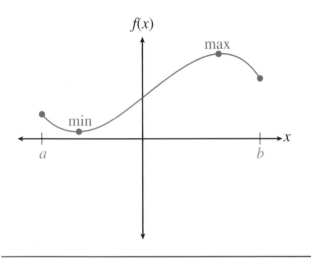

Case II: *f* has an extreme value at *a* and another extreme value in (*a,b*).

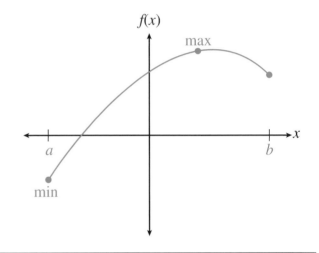

Case III: *f* has an extreme value at *b* and another extreme value in (*a,b*).

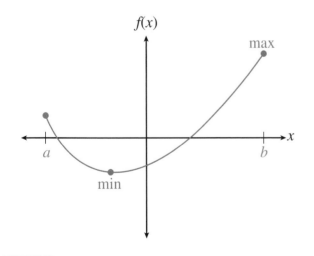

Case IV: *f* has the same maximum and minimum value.

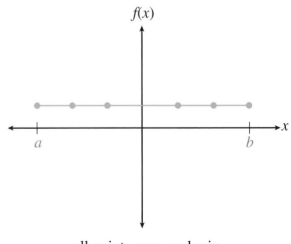

all points max. and min.

2

Algebraic Methods to Calculate Limits

This chapter presents a variety of algebraic methods for calculating limits. Where indeterminate forms occur, techniques are introduced, such as factor/reduce, dividing by the largest power of the variable, rationalizing the denominator/numerator, and finding the least common denominator. The chapter ends with locating horizontal asymptotes for the graph of a function, a process that involves finding limits at infinity.

Direct Substitution 36

Indeterminate Forms $\pm\frac{\infty}{\infty}$ and $\frac{0}{0}$ 38

Dealing with Indeterminate Forms . . . 39

Limits at Infinity: Horizontal
    Asymptotes 48

Direct Substitution

Many times, you can find the limit $\lim_{x \to c} f(x)$ by simply substituting c for x and then evaluating the resulting expression. In this section, you will see examples of how to determine the limits of polynomial, radical, and trigonometric functions, as well as how to determine the limit of a quotient of rational expressions.

LIMIT OF A POLYNOMIAL FUNCTION

Determine $\lim_{x \to 2} (x^2 + 3x)$

❶ Begin with the original limit statement.

$$\lim_{x \to 2} (x^2 + 3x)$$

❷ Substitute 2 for x, then simplify.

$$= 2^2 + 3(2)$$
$$= 4 + 6 = 10$$

LIMIT INVOLVING A RADICAL FUNCTION

Determine $\lim_{x \to 5} \dfrac{\sqrt{x-1}}{3x}$.

❶ Begin with the original limit statement.

$$\lim_{x \to 5} \frac{\sqrt{x-1}}{3x}$$

❷ Substitute 5 for x, then simplify.

$$= \frac{\sqrt{5-1}}{3x}$$
$$= \frac{\sqrt{4}}{15} = \frac{2}{5}$$

LIMIT INVOLVING A TRIGONOMETRIC FUNCTION

Determine $\lim\limits_{x \to \pi} \dfrac{\cos x}{2x}$.

❶ Begin with the original limit statement.

$$\lim\limits_{x \to \pi} \dfrac{\cos x}{2x}$$

❷ Substitute π for x; then simplify.

$$= \dfrac{\cos \pi}{2 \cdot \pi}$$

$$= \dfrac{-1}{2\pi}$$

LIMIT OF A QUOTIENT OF RATIONAL EXPRESSIONS

Determine $\lim\limits_{x \to \infty} \dfrac{5 + \dfrac{3}{x} + \dfrac{7}{x^2}}{7 - \dfrac{4}{x} + \dfrac{6}{x^2}}$.

❶ Begin with the original limit statement.

$$\lim\limits_{x \to \infty} \dfrac{5 + \dfrac{3}{x} + \dfrac{7}{x^2}}{7 - \dfrac{4}{x} + \dfrac{6}{x^2}}$$

❷ Substitute ∞ for x; then simplify.

$$= \dfrac{5 + 0 + 0}{7 - 0 + 0}$$

$$= \dfrac{5}{7}$$

Indeterminate Forms $\pm\frac{\infty}{\infty}$ and $\frac{0}{0}$

Sometimes when using direct substitution to calculate a limit, you may encounter expressions such as $\pm\frac{\infty}{\infty}$ or $\frac{0}{0}$. These are known as indeterminate forms. When you encounter indeterminate forms, appropriate algebraic methods must be used to alter the form of the expression the limit of which you are attempting to calculate.

INDETERMINATE FORM INVOLVING TRIGONOMETRIC FUNCTION

Here, x is being replaced with the number that x is approaching—it's what substitution is all about. Meanwhile, the colors show that an appropriate number is being substituted for the x.

Determine $\lim\limits_{x \to 0} \dfrac{\sin 3x}{x}$.

$$\lim_{x \to 0} \frac{\sin 3x}{x}$$
$$= \frac{\sin(3 \cdot 0)}{0}$$
$$= \frac{\sin 0}{0}$$
$$= \frac{0}{0}$$

INDETERMINATE FORM INVOLVING RATIONAL FUNCTION

Determine $\lim\limits_{x \to \infty} \dfrac{x^2 + 3x}{7 - 2x^2}$.

$$\lim_{x \to \infty} \frac{x^2 + 3x}{7 - 2x^2}$$
$$= \frac{\infty^2 + 3 \cdot \infty}{7 - 2 \cdot \infty^2}$$
$$= \frac{\infty}{-\infty}$$

INDETERMINATE FORM INVOLVING RECIPROCALS

Determine $\lim\limits_{x \to \infty} \dfrac{\dfrac{1}{x} + \dfrac{2}{x+1}}{\dfrac{3}{x^2 + x}}$.

$$\lim_{x \to \infty} \frac{\dfrac{1}{x} + \dfrac{2}{x+1}}{\dfrac{3}{x^2 + x}}$$
$$= \frac{\dfrac{1}{\infty} + \dfrac{2}{\infty + 1}}{\dfrac{3}{\infty^2 + \infty}}$$
$$= \frac{0 + 0}{0}$$
$$= \frac{0}{0}$$

Dealing with Indeterminate Forms

When you encounter an indeterminate form, you can use a variety of algebraic techniques to determine the limits. Among these techniques are factoring and reducing, dividing by the largest power of the variable, using the common denominator, and rationalizing the denominator (or the numerator).

Factor and Reduce

Using this technique, you factor the numerator and denominator, cancel like factors, and then use direct substitution to evaluate the resulting expression.

LIMIT OF A RATIONAL FUNCTION

Determine $\lim\limits_{x \to \infty} \dfrac{x^2 - 5x + 6}{x^2 + 3x - 10}$.

1 Try direct substitution, 2 for x.

$$\lim\limits_{x \to 2} \dfrac{x^2 - 5x + 6}{x^2 + 3x - 10}$$

$$= \dfrac{2_2 - 5 \cdot 2 + 6}{2^2 + 3 \cdot 2 - 10}$$

$$= \dfrac{4 - 10 + 6}{4 + 6 - 10}$$

$$= \dfrac{0}{0}$$

2 Since you ended up with an *indeterminate form*, return to the original limit statement and then factor both the numerator and the denominator, cancel the common factor, and then use direct substitution.

$$\lim\limits_{x \to 2} \dfrac{x^2 - 5x + 6}{x^2 + 3x - 10}$$

$$= \lim\limits_{x \to 2} \dfrac{(x - 3)(x - 2)}{(x + 5)(x - 2)}$$

$$= \lim\limits_{x \to 2} \dfrac{x - 3}{x + 5}$$

$$= \dfrac{2 - 3}{2 + 5}$$

$$= \dfrac{-1}{7}$$

Dealing with Indeterminate Forms *(continued)*

LIMIT OF A RATIO OF TRIGONOMETRIC FUNCTIONS

Determine $\lim\limits_{x \to 0} \dfrac{\sin x + \sin 2x}{\sin x}$.

❶ Try direct substitution, 0 for x.

$$\lim_{x \to 0} \frac{\sin x + \sin 2x}{\sin x}$$
$$= \frac{\sin(0) + \sin 2(0)}{\sin(0)}$$
$$= \frac{0 + 0}{0}$$
$$= \frac{0}{0}$$

❷ Again, you ended up with an indeterminate form. Return to the original limit statement and substitute $2\sin x \cos x$ *for* $\sin 2x$. Then factor, reduce, and use direct substitution.

$$\lim_{x \to 0} \frac{\sin x + \sin 2x}{\sin x}$$
$$= \lim_{x \to 0} \frac{\sin x + 2\sin x \cos x}{\sin x}$$
$$= \lim_{x \to 0} \frac{\sin x(1 + 2\cos x)}{\sin x}$$
$$= \lim_{x \to 0} (1 + 2\cos x)$$
$$= 1 + 2\cos(0)$$
$$= 1 + 2 \cdot 1$$
$$= 3$$

Divide by Largest Power of the Variable

When the limit involves a rational function, you can divide all terms by the highest power of the variable in the rational function—or you can multiply by the reciprocal of the highest powered term instead.

LIMIT OF A RATIONAL FUNCTION

Determine $\lim\limits_{x \to \infty} \dfrac{5x^2 + 3x - 2}{7x^2 - 4x + 6}$

❶ Try direct substitution, ∞ for x.

$$\lim_{x \to \infty} \frac{5x^2 + 3x - 2}{7x^2 - 4x + 6}$$

$$= \frac{5(\infty)^2 + 3(\infty) - 2}{7(\infty)^2 - 4(\infty) + 6}$$

$$= \frac{\infty}{\infty}$$

❷ Return to the original limit statement and multiply the numerator and denominator by $\frac{1}{x^2}$, x^2 being the highest powered variable term.

$$\lim_{x \to \infty} \frac{(5x^2 + 3x - 2)}{(7x^2 - 4x + 6)} \cdot \frac{\frac{1}{x^2}}{\frac{1}{x^2}}$$

$$= \lim_{x \to \infty} \frac{\frac{5x^2}{x^2} + \frac{3x}{x^2} - \frac{2}{x^2}}{\frac{7x^2}{x^2} - \frac{4x}{x^2} + \frac{2}{x^2}}$$

$$= \lim_{x \to \infty} \frac{5 + \frac{3}{x} - \frac{2}{x^2}}{7 - \frac{4}{x} + \frac{6}{x^2}}$$

$$= \frac{5 + \frac{3}{\infty} - \frac{2}{\infty^2}}{7 - \frac{4}{\infty} + \frac{6}{\infty^2}}$$

$$= \frac{5 + 0 - 0}{7 - 0 + 0}$$

$$= \frac{5}{7}$$

LIMIT INVOLVING A RADICAL FUNCTION

Determine $\lim\limits_{x \to \infty} \dfrac{\sqrt{x^2 - 2x}}{5x - 3}$

❶ Try direct substitution, ∞ for x.

$$\lim_{x \to \infty} \frac{\sqrt{x^2 - 2x}}{5x - 3}$$

$$= \frac{\sqrt{\infty^2 - 2(\infty)}}{5(\infty) - 3}$$

$$= \frac{\infty}{\infty}$$

❷ Since you got another indeterminate form, multiply the numerator and the denominator by $\dfrac{1}{\sqrt{x^2}}$.

$$\lim_{x \to \infty} \frac{\sqrt{x^2 - 2x}}{5x - 3} \cdot \frac{\frac{1}{\sqrt{x^2}}}{\frac{1}{\sqrt{x^2}}}$$

$$= \lim_{x \to \infty} \frac{\sqrt{\dfrac{x^2 - 2x}{x^2}}}{\dfrac{5x - 3}{\sqrt{x^2}}}$$

> **TIP**
>
> Use the largest power of the variable x in whatever form it appears.
>
> For $\sqrt{x^2 - 2x}$, use $\dfrac{1}{\sqrt{x^2}}$.
>
> For $\sqrt{5x + x^3}$, use $\dfrac{1}{\sqrt{x^3}}$.

$$= \lim_{x \to \infty} \frac{\sqrt{\dfrac{x^2}{x^2} - \dfrac{2x}{x^2}}}{\dfrac{5x}{x} - \dfrac{3}{x}}$$

$$= \lim_{x \to \infty} \frac{\sqrt{1 - \dfrac{2}{x}}}{5 - \dfrac{3}{x}}$$

$$= \frac{\sqrt{1 - \dfrac{2}{\infty}}}{5 - \dfrac{3}{\infty}}$$

$$= \frac{\sqrt{1 - 0}}{5 - 0}$$

$$= \frac{1}{5}$$

Use the Common Denominator

When an expression involves rational terms, you use the least common denominator of all rational terms.

LIMIT INVOLVING RATIONAL EXPRESSIONS: EXAMPLE 1

Determine $\lim\limits_{x \to -3} \dfrac{\frac{1}{x} + \frac{1}{3}}{x + 3}$

1 Try direct substitution, -3 for x.

$$\lim\limits_{x \to -3} \frac{\frac{1}{x} + \frac{1}{3}}{x + 3}$$

$$= \frac{\frac{1}{-3} + \frac{1}{3}}{-3 + 3}$$

$$= \frac{0}{0}$$

2 Since you encountered an indeterminate form, return to the original limit. The least common denominator for all fractions is $3x$. Change both fractions in the numerator to this common denominator.

Note: Multiplying all terms in the numerator and denominator by $3x$ gives you the same result that appears in Step 3 (see following page.)

$$\lim\limits_{x \to -3} \frac{\frac{1}{x} + \frac{1}{3}}{x + 3}$$

$$= \lim\limits_{x \to -3} \frac{\frac{1}{x}\left(\frac{3}{3}\right) + \frac{1}{3}\left(\frac{x}{x}\right)}{x + 3}$$

$$= \lim\limits_{x \to -3} \frac{\frac{3}{3x} + \frac{x}{3x}}{x + 3}$$

$$= \lim\limits_{x \to -3} \frac{\frac{3 + x}{3x}}{x + 3}$$

❸ Invert and multiply, simplify the fraction, and then use direct substitution, −3 for x.

$$=\lim_{x\to-3}\frac{3+x}{3x}\cdot\frac{1}{x+3}$$

$$=\lim_{x\to-3}\frac{1}{3x}$$

$$=\frac{1}{3(-3)}$$

$$=-\frac{1}{9}$$

LIMIT INVOLVING RATIONAL EXPRESSIONS: EXAMPLE 2

Determine $\displaystyle\lim_{x\to\infty}\frac{\dfrac{1}{x}+\dfrac{2}{x+1}}{\dfrac{3}{x^2+x}}$

❶ Try direct substitution, ∞ for x.

$$\lim_{x\to\infty}\frac{\dfrac{1}{x}+\dfrac{2}{x+1}}{\dfrac{3}{x^2+x}}$$

$$=\frac{\dfrac{1}{\infty}+\dfrac{2}{\infty+1}}{\dfrac{3}{\infty^2+\infty}}$$

$$=\frac{0+0}{0}$$

$$=\frac{0}{0}$$

> **TIP**
>
> $\dfrac{1}{\infty}$ approaches 0, as do $\dfrac{1}{\infty+1}$ and $\dfrac{1}{\infty^2+\infty}$.

❷ Having encountered an indeterminate form, return to the original limit statement, writing all fractions in terms of the least common denominator of all of the terms: $x(x + 1)$.

Note: The goal is to eventually get rid of all of the denominators in both the top and bottom of the original fraction.

$$\lim_{x \to \infty} \frac{\dfrac{1}{x} + \dfrac{2}{x+1}}{x^2 + x}$$

$$= \lim_{x \to \infty} \frac{\dfrac{1}{x} + \dfrac{2}{x+1}}{\dfrac{3}{x(x+1)}}$$

$$= \lim_{x \to \infty} \frac{\dfrac{1}{x} \cdot \dfrac{x+1}{x+1} + \dfrac{2}{x+1} \cdot \dfrac{x}{x}}{\dfrac{3}{x(x+1)}}$$

$$= \lim_{x \to \infty} \frac{\dfrac{x+1}{x(x+1)} + \dfrac{2x}{x(x+1)}}{\dfrac{3}{x(x+1)}}$$

$$= \lim_{x \to \infty} \frac{\dfrac{3x+1}{x(x+1)}}{\dfrac{3}{x(x+1)}}$$

❸ Invert and multiply, then simplify the resulting expression.

$$= \lim_{x \to \infty} \frac{3x+1}{x(x+1)} \cdot \frac{x(x+1)}{3}$$

$$= \lim_{x \to \infty} \frac{3x+1}{3}$$

$$= \frac{3(\infty) + 1}{3}$$

$$= \infty$$

Therefore, limit does not exist.

Dealing with Indeterminate Forms *(continued)*

Rationalize the Numerator (or Denominator)

In this technique, use the conjugate of a radical expression to calculate the limit.

RATIONALIZE THE NUMERATOR

Calculate $\lim\limits_{x \to 9} \dfrac{\sqrt{x} - 3}{x - 9}$.

① Try direct substitution, 9 for x; an indeterminate form results.

$$\lim\limits_{x \to 9} \frac{\sqrt{x} - 3}{x - 9}$$
$$= \frac{\sqrt{9} - 3}{9 - 9}$$
$$= \frac{0}{0}$$

② Multiply the numerator and denominator by $\sqrt{x} + 3$, the conjugate of $\sqrt{x} - 3$, simplify the resulting expression, and then use direct substitution.

$$\lim\limits_{x \to 9} \frac{\sqrt{x} - 3}{x - 9} \cdot \frac{\sqrt{x} + 3}{\sqrt{x} + 3}$$
$$= \lim\limits_{x \to 9} \frac{x - 9}{(x - 9)(\sqrt{x} + 3)}$$
$$= \lim\limits_{x \to 9} \frac{1}{\sqrt{x} + 3}$$
$$= \frac{1}{\sqrt{9} + 3}$$
$$= \frac{1}{3 + 3} = \frac{1}{6}$$

> **TIP**
>
> $\left(\sqrt{\text{1st term}} - \text{number}\right)\left(\sqrt{\text{1st term}} + \text{number}\right)$
> just equals 1st term $+ (\text{number})^2$.

RATIONALIZE THE DENOMINATOR

Calculate $\lim\limits_{x \to 2} \dfrac{x-2}{\sqrt{x+3} - \sqrt{5}}$

❶ Direct substitution, 2 for x, leads to an indeterminate form.

$$\lim_{x \to 2} \frac{x-2}{\sqrt{x+3} - \sqrt{5}}$$

$$= \frac{2-2}{\sqrt{2+3} - \sqrt{5}}$$

$$= \frac{0}{0}$$

❷ Multiply the numerator and denominator of the original expression by $\sqrt{x+3} + \sqrt{5}$, the conjugate of $\sqrt{x+3} - \sqrt{5}$. Then simplify the result.

$$\lim_{x \to 2} \frac{x-2}{\sqrt{x+3} - \sqrt{5}} \cdot \frac{\sqrt{x+3} + \sqrt{5}}{\sqrt{x+3} + \sqrt{5}}$$

$$= \lim_{x \to 2} \frac{(x-2)(\sqrt{x+3} + \sqrt{5})}{(x+3) - 5}$$

$$= \lim_{x \to 2} \frac{(x-2)(\sqrt{x+3} + \sqrt{5})}{(x-2)}$$

❸ Use direct substitution in the resulting expression.

$$= \lim_{x \to 2} \left(\sqrt{x+3} + \sqrt{5}\right)$$

$$= \sqrt{2+3} + \sqrt{5}$$

$$= \sqrt{5} + \sqrt{5}$$

$$= 2\sqrt{5}$$

Limits at Infinity: Horizontal Asymptotes

This section discusses the behavior of the graph of a function as x approaches $\pm\infty$, in other words, limits at infinity.

Definition of a Horizontal Asymptote

The line $y = L$ is a **horizontal asymptote** of the graph of $y = f(x)$ if either

$\lim\limits_{x \to \infty} f(x) = L$ or $\lim\limits_{x \to \infty} f(x) = L$. Infrequently,

$\lim\limits_{x \to \infty} f(x)$ and $\lim\limits_{x \to \infty} f(x)$ are not the same number; in which case there can be two different horizontal asymptotes (see the fourth example at the right.)

In the first figure, $\lim\limits_{x \to -\infty} = L$.

In the second figure, $\lim\limits_{x \to \pm\infty} = L$.

In the third figure, $\lim\limits_{x \to -\infty} = L$.

In the fourth figure, $\lim\limits_{x \to -\infty} = M$, but $\lim\limits_{x \to +\infty} = L$.

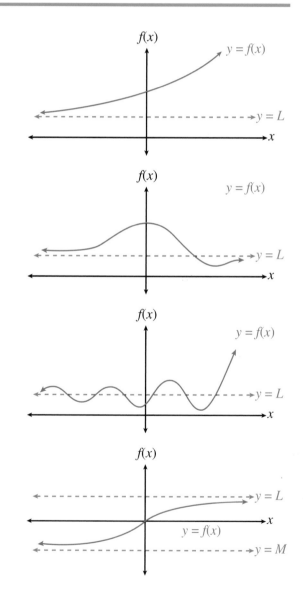

A FUNCTION WITH THE *X*-AXIS AS ITS HORIZONTAL ASYMPTOTE

Find the horizontal asymptote for the graph of $f(x) = \dfrac{3}{x-1}$.

❶ Set up the limit statement as x approaches ∞ and then evaluate the limit.

$$\lim_{x \to \infty} \frac{3}{x-1}$$
$$= \frac{3}{\infty - 1}$$
$$= 0$$

❷ At the right is the graph of $f(x) = \dfrac{3}{x-1}$, with

its horizontal asymptote at $y = 0$. Note that the

graph also has a vertical asymptote at $x = 1$, the

point at which the function is undefined, i.e., its

denominator equals 0.

Note: *Substituting* $-\infty$ *for x would have given you the same result of y = 0.*

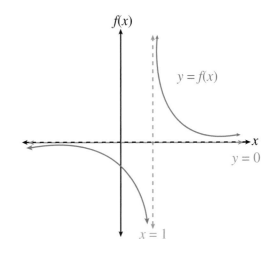

HORIZONTAL ASYMPTOTE OF A RATIONAL FUNCTION

Find the horizontal asymptote for the graph of $f(x) = \dfrac{5x^2 - 3x + 2}{7x^2 + 7x - 42}$.

❶ Set up the limit statement as x approaches ∞.

$$\lim_{x \to \infty} \frac{5x^2 - 3x + 2}{7x^2 + 7x - 42}$$

② If you were to use direct substitution, an indeterminate form would result. Instead, divide all terms by x^2, then simplify.

$$\lim_{x \to \infty} \frac{5x^2 - 3x + 2}{7x^2 + 7x - 42}$$

$$= \lim_{x \to \infty} \frac{\dfrac{5x^2}{x^2} - \dfrac{3x}{x^2} + \dfrac{2}{x^2}}{\dfrac{7x^2}{x^2} + \dfrac{7x}{x^2} - \dfrac{42}{x^2}}$$

$$= \lim_{x \to \infty} \frac{5 - \dfrac{3}{x} + \dfrac{2}{x^2}}{7 + \dfrac{7}{x} - \dfrac{42}{x^2}}$$

③ Now substitute ∞ for x and simplify the result.

$$= \frac{5 - \dfrac{3}{\infty} + \dfrac{2}{\infty^2}}{7 + \dfrac{7}{\infty} - \dfrac{42}{\infty^2}}$$

$$= \frac{5 - 0 + 0}{7 + 0 - 0}$$

$$= \frac{5}{7}; \text{ therefore } y = \frac{5}{7} \text{ is the horizontal asymptote.}$$

Note: *Substituting* $-\infty$ *for x would have given you the same result of* $y = \dfrac{5}{7}.$

4 At the right is the graph of

$f(x) = \dfrac{5x^2 - 3x + 2}{7x^2 + 7x - 42}$ with its horizontal

asymptote at $y = \dfrac{5}{7}$, along with its two

vertical asymptotes at $x = -3$ and $x = 2$.

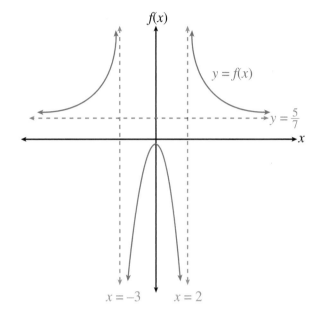

HORIZONTAL ASYMPTOTE OF A RATIONAL FUNCTION USING MULTIPLE LIMIT TECHNIQUES

Find the horizontal asymptote for the graph of $f(x) = \dfrac{2x^2 - x - 6}{x^2 + 2x - 8}$.

1 Set up the limit statement as x approaches ∞.

$$\lim_{x \to \infty} \frac{2x^2 - x - 6}{x^2 + 2x - 8}$$

2 If you used direct substitution, you would encounter an indeterminate form. Next factor the numerator and denominator and then reduce the resulting expression.

$$\lim_{x \to \infty} \frac{2x^2 - x - 6}{x^2 + 2x - 8}$$

$$= \lim_{x \to \infty} \frac{(2x + 3)(x - 2)}{(x + 4)(x - 2)}$$

$$= \lim_{x \to \infty} \frac{2x + 3}{x + 4}$$

3 If you were to directly substitute at this point, an indeterminate form again would result. Instead, divide all terms by x, and then simplify.

$$=\lim_{x \to \infty} \frac{\frac{2x}{x} + \frac{3}{x}}{\frac{x}{x} + \frac{4}{x}}$$

$$=\lim_{x \to \infty} \frac{2 + \frac{3}{x}}{1 + \frac{4}{x}}$$

4 Now use direct substitution, ∞ for x.

Note: *Substituting $-\infty$ for x would have given us the same result of y = 2.*

$$=\frac{2 + \frac{3}{\infty}}{1 + \frac{4}{\infty}}$$

$$=\frac{2 + 0}{1 + 0}$$

$$=2$$

5 The graph of $f(x) = \dfrac{2x^2 - x - 6}{x^2 + 2x - 8}$ is shown

at the right with its horizontal asymptote at

$y = 2$.

Note the "hole" in the graph at $x = 2$. In Step 2 (see preceding page), the factor $x - 2$ was cancelled, so $x \neq 2$.

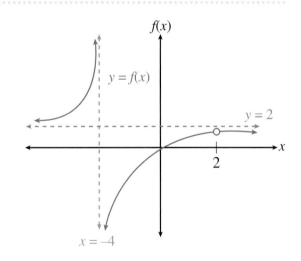

FUNCTION WHOSE GRAPH HAS 2 HORIZONTAL ASYMPTOTES

Find the horizontal asymptote(s) for the graph of $f(x) = \dfrac{3x+2}{\sqrt{x^2+3}-1}$.

1 Set up the limit statement as x approaches ∞.

$$\lim_{x \to \infty} \frac{3x+2}{\sqrt{x^2+3}-1}$$

2 Direct substitution would result in an indeterminate form. Divide all terms by x and then simplify where possible.

$$\lim_{x \to \infty} \frac{\dfrac{3x}{x}+\dfrac{2}{x}}{\dfrac{\sqrt{x^2+3}}{x}-1}$$

$$=\lim_{x \to \infty} \frac{3+\dfrac{2}{x}}{\sqrt{\dfrac{x^2+3}{x^2}}-\dfrac{1}{x}}$$

$$=\lim_{x \to \infty} \frac{3+\dfrac{2}{x}}{\sqrt{1+\dfrac{3}{x^2}}-\dfrac{1}{x}}$$

3 Last, substitute ∞ for and simplify.

$$=\frac{3+\dfrac{2}{\infty}}{\sqrt{1+\dfrac{3}{\infty^2}}-\dfrac{1}{\infty}}$$

$$=\frac{3+0}{\sqrt{1+0}-0}$$

$$=3$$

So, $y = 3$ is a horizontal asymptote.

❹ Let's take another look at the original function.

$$f(x) = \frac{3x+2}{\sqrt{x^2+3}-1}$$

a) $x \to +\infty$

$f(x) = \dfrac{positive\ number}{positive\ number}$ so the

horizontal asymptote is $y = 3$.

b) but as $x \to -\infty$

$f(x) = \dfrac{negative\ number}{positive\ number}$ so the

horizontal asymptote is $y = -3$.

❺ At the right is the graph of $f(x) = \dfrac{3x+2}{\sqrt{x^2+3}-1}$ with its

two horizontal asymptotes, $y = 3$, and $y = -3$.

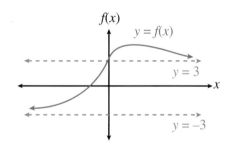

chapter 3

Introduction to the Derivative

us is
nding a
mmon uses
 the formal
es. The
ations for the
plications
ter concludes
tiability and

What Can Be Done With
a Derivative?. 56

Derivative as the Slope of
a Tangent Line 58

Derivative by Definition. 60

Find the Equation of a Line
Tangent to a Curve. 67

Horizontal Tangents. 68

Alternate Notations for a Derivative . . . 70

Derivative as a Rate of Change. 72

Differentiability and Continuity 74

This section introduces some common uses for a derivative.

FIND SLOPE OF A TANGENT LINE

You can use a derivative to find the slope of a line tangent to the graph of a function at a given point P.

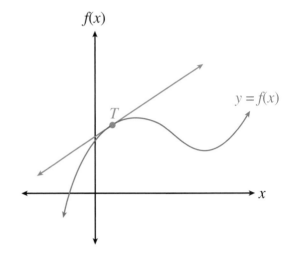

MAXIMUM AND MINIMUM ON GRAPH OF A FUNCTION

You can also use a derivative to find points on the graph of a function where the relative maximum and relative minimum occur.

Intervals on the graph that are increasing or decreasing can also be found.

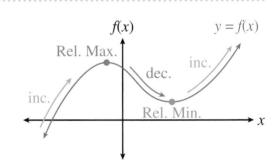

ANALYZE RATES OF CHANGE

You can use a derivative to analyze rates of change. Given the formula for the volume of a sphere, you can use the derivative to relate the rate of change of the volume, $\frac{dV}{dt}$, to the rate of change of its radius, $\frac{dr}{dt}$.

$$V_{sphere} = \frac{4}{3}\pi r^3$$

$$\frac{dV}{dt} = 4\pi r^2 \frac{dr}{dt}$$

ANALYZE MOTION ON A OBJECT

Given a function, $s(t)$, that describes the position of an object, you can use derivatives to find both the velocity function, $v(t)$, and the acceleration function, $a(t)$.

$$s(t) = t^3 - 5t^2 + 7t - 15$$
$$v(t) = 3t^2 - 10t + 7$$
$$a(t) = 6t - 10$$

OPTIMIZE WORD PROBLEMS

Let's say that equal squares are cut from each corner of a rectangular sheet of metal which is 10 inches by 6 inches. After removing the squares at each corner, the "flaps" are folded up to create a box with no top.

You can use a derivative to find the size of each square to be removed so that the resulting box has the maximum volume.

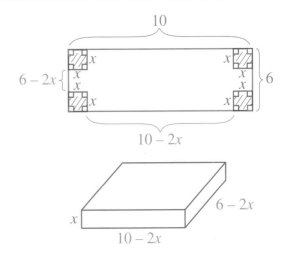

Volume Box $= x\,(10 - 2x)(6 - 2x)$

Find the slope of the line tangent to the graph of $f(x) = x^2 - 3x$ at the point with x-coordinate 2.

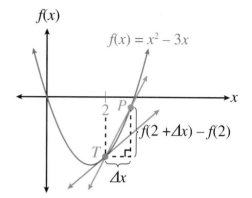

$f(x)$

$f(x) = x^2 - 3x$

$f(2 + \Delta x) - f(2)$

Δx

❶ Select a point P that is Δx units to the right of point T.

❷ The coordinates of point T are $\left(2, f(2)\right)$.

The coordinates of point P are $(\ 2 + \Delta x, f(2 + \Delta x)\ )$

❸ Find the slope of the secant line $\overleftrightarrow{TP}$.

$$slope\ of\ \overleftrightarrow{TP} = \frac{y_2 - y_1}{x_2 - x_1}$$

$$= \frac{f(2 + \Delta x) - f(2)}{(2 + \Delta x) - 2}$$

❹ As $\Delta x \to 0$, the slope of the secant line $\overleftrightarrow{TP}$ gets closer and closer to the line tangent at $x = 2$.

The slope of the tangent line at point $T = \lim\limits_{\Delta x \to 0} \dfrac{f(2 + \Delta x) - f(2)}{(2 + \Delta x) - 2}$.

5 Find $f(2 + \Delta x)$ and $f(2)$ using $f(x) = x^2 - 3x$, then simplify.

$$= \lim_{\Delta x \to 0} \frac{\left[(2 + \Delta x)^2 - 3(2 + \Delta x)\right] - \left[2^2 - 3 \cdot 2\right]}{2 + \Delta x - 2}$$

$$= \lim_{\Delta x \to 0} \frac{4 + 4 \cdot \Delta x + (\Delta x)^2 - 6 - 3\Delta x - 4 + 6}{\Delta x}$$

$$= \lim_{\Delta x \to 0} \frac{\Delta x + (\Delta x)^2}{\Delta x}$$

6 Factor out Δx, cancel like factors, then substitute 0 for Δx.

$$= \lim_{\Delta x \to 0} \frac{\cancel{\Delta x}(1 + \Delta x)}{\cancel{\Delta x}}$$

$$= \lim_{\Delta x \to 0} (1 + \Delta x)$$

$$= 1 + 0$$

$$= 1$$

7 The last expression, 1, above on the right, is the slope of the line tangent to the graph of $f(x) = x^2 - 3x$ at the point where $x = 2$.

Therefore, the slope of line tangent at $x = 2$ is 1.

The limit process used above to find the slope of the line tangent at $x = 2$ is called the **derivative of $f(x)$ at $x = 2$,** denoted as $f'(\mathbf{2})$ (read "f prime of 2").

$$f'(2) = \lim_{\Delta x \to 0} \frac{f(2 + \Delta x) - f(2)}{\Delta x}$$

Derivative by Definition

As you can see from the previous example, when you replace the Δx with an h and the 2 with an arbitrary number c, you end with a formal definition of the derivative of $f(x)$ at $x = c$.

FIRST FORM OF THE DEFINITION

This is the first of the two most common definition forms.

The *derivative of a function f at a number c*, denoted by $f'(c)$, is given by the statement to the right.

$$f'(c) = \lim_{h \to 0} \frac{f(c+h) - f(c)}{h}$$

As $h \to 0$, point P gets closer and closer to point T.

Note: *The process of finding a derivative is called* **differentiation.**

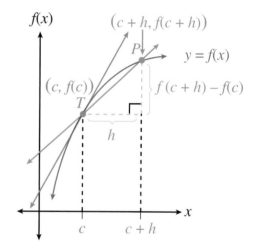

$$\text{slope of } \overleftrightarrow{TP} = \frac{f(c+h) - f(c)}{(c+h) - c}$$

$$= \frac{f(c+h) - f(c)}{h}$$

$$\text{slope of line tangent at point } T = \lim_{h \to 0} \frac{f(c+h) - f(c)}{h} = f'(c)$$

Determine the Derivative of a Specific Function at a Specific Number

USING THE FIRST FORM OF THE DEFINITION

Using the definition above, find the derivative of $f(x) = x^2 - 5x + 3$ at $x = 2$; that is, find $f'(2)$.

---

❶ Set up the limit statement from the definition.

$$f'(2) = \lim_{h \to 0} \frac{f(2+h) - f(2)}{h}$$

---

❷ Find $f(2+h)$ and $f(2)$ by using $f(x) = x^2 - 5x + 3$.

$$= \lim_{h \to 0} \frac{\left[(2+h)^2 - 5(2+h) + 3\right] - \left[2^2 - 5 \cdot 2 + 3\right]}{h}$$

---

❸ Expand the numerator and simplify the resulting expression.

Therefore, $f'(2) = -1$

Note: −1 is actually the slope of the line tangent to the graph of $f(x) = x^2 - 5x + 3$ at the point with x-coordinate 2.

$$= \lim_{h \to 0} \frac{4 + 4h + h^2 - 10 - 5h + 3 - 4 + 10 - 3}{h}$$

$$= \lim_{h \to 0} \frac{-h + h^2}{h}$$

$$= \lim_{h \to 0} \frac{\cancel{h}(-1 + h)}{\cancel{h}}$$

$$= \lim_{h \to 0} (-1 + h)$$

$$= -1$$

SECOND FORM OF THE DEFINITION

The *derivative of a function f at a number c*, denoted by
$f'(c)$ is given by:

$$f'(c) = \lim_{x \to c} \frac{f(x) - f(c)}{x - c}$$

As $x \to c$, the point P gets closer and closer to point T.

$$slope \ of \ \overleftrightarrow{TP} = \frac{f(x) - f(c)}{x - c}$$

Slope of line tangent at point T

$$= \lim_{x \to c} \frac{f(x) - f(c)}{x - c} = f'(c)$$

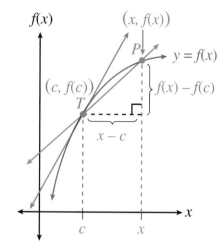

Find the Derivative of a Specific Function at a Specific Number

USING THE SECOND FORM OF THE DEFINITION

Using the definition above, find the derivative of $f(x) = x^2 - 5x + 3$ at $x = 2$, that is, find $f'(2)$.

❶ Set up the limit statement from the second form of the
derivative definition.

$$f'(2) = \lim_{x \to 2} \frac{f(x) - f(2)}{x - 2}$$

2 Find $f(x)$ and $f(2)$ using $f(x) = x^2 - 5x + 3$, and then simplify.

$$= \lim_{x \to 2} \frac{\left[x^2 - 5x + 3\right] - \left[2^2 - 5 \cdot 2 + 3\right]}{x - 2}$$

$$= \lim_{x \to 2} \frac{x^2 - 5x + 3 - 4 + 10 - 3}{x - 2}$$

$$= \lim_{x \to 2} \frac{x^2 - 5x + 6}{x - 2}$$

3 Direct substitution leads to the indeterminate form 0/0. Instead, factor the numerator, cancel the common factor, and then use direct substitution.

Therefore, $f'(2) = -1$

Note: *This is the same result as finding $f'(2)$ using the first form of the definition of the derivative.*

$$= \lim_{x \to 2} \frac{(x - 2)(x - 3)}{x - 2}$$

$$= \lim_{x \to 2} (x - 3)$$

$$= 2 - 3$$

$$= -1$$

DERIVATIVE OF A SPECIFIC POLYNOMIAL FUNCTION

For $f(x) = 3x^2 - 12x + 9$, find $f'(x)$, the derivative at any point.

1 Set up the limit process.

$$f'(c) = \lim_{h \to 0} \frac{f(x + h) - f(x)}{h}$$

2 Find $f(x + h)$ and $f(x)$ using $f(x) = 3x^2 - 12x + 9$, expand the numerator, and then simplify.

$$\lim_{h \to 0} \frac{\left[3(x + h)^2 - 12(x + h) + 9\right] - \left[3x^2 - 12x + 9\right]}{h}$$

$$= \lim_{h \to 0} \frac{3x^2 + 6xh + 3h^2 - 12x - 12h + 9 - 3x^2 + 12x - 9}{h}$$

$$= \lim_{h \to 0} \frac{6xh + 3h^2 - 12h}{h}$$

❸ Factor out h in the numerator, cancel like factors, and then finish up with direct substitution.

$$=\lim_{h\to 0}\frac{\cancel{h}(6x+3h-12)}{\cancel{h}}$$

$$=\lim_{h\to 0}(6x+3h-12)$$

$$=6x+3\cdot 0-12$$

Therefore $f'(x) = 6x - 12$

$$=6x-12$$

DERIVATIVE OF A RADICAL FUNCTION

Find $f'(x)$ for $f(x) = \sqrt{x+3}$

❶ Begin with the limit statement from the derivative definition.

$$f'(x) = \lim_{h\to 0}\frac{f(x+h)-f(x)}{h}$$

❷ Using $f(x) = \sqrt{x+3}$, find $f(x+h)$ and $f(x)$.

$$=\lim_{h\to 0}\frac{\sqrt{x+h+3}-\sqrt{x+3}}{h}$$

❸ Direct substitution leads to an indeterminate form. In this case, multiply the numerator and denominator by $\sqrt{x+h+3}+\sqrt{x+3}$, the conjugate of the numerator.

$$=\lim_{h\to 0}\frac{\sqrt{x+h+3}-\sqrt{x+3}}{h}\cdot\frac{\sqrt{x+h+3}+\sqrt{x+3}}{\sqrt{x+h+3}+\sqrt{x+3}}$$

④ Expand the numerator and simplify it, but leave the denominator as is.

$$=\lim_{h \to 0} \frac{(x+h+3)-(x+3)}{h\left(\sqrt{x+h+3}+\sqrt{x+3}\right)}$$

$$=\lim_{h \to 0} \frac{x+h+3-x-3}{h\left(\sqrt{x+h+3}+\sqrt{x+3}\right)}$$

TIP

The product,
$$\left(\sqrt{\text{1st term}}-\sqrt{\text{2nd term}}\right)\left(\sqrt{\text{1st term}}+\sqrt{\text{2nd term}}\right),$$
is just 1st term − 2nd term.

$$=\lim_{h \to 0} \frac{\cancel{h}}{\cancel{h}\left(\sqrt{x+h+3}+\sqrt{x+3}\right)}$$

$$=\lim_{h \to 0} \frac{1}{\sqrt{x+h+3}+\sqrt{x+3}}$$

⑤ Last, use direct substitution, 0 for h.

$$=\frac{1}{\sqrt{x+0+3}+\sqrt{x+3}}$$

$$=\frac{1}{\sqrt{x+3}+\sqrt{x+3}}$$

$$=\frac{1}{2\sqrt{x+3}}$$

Therefore, $f'(x)=\dfrac{1}{2\sqrt{x+3}}$

TIP

Definition:
If the derivative of a function (or the derivative at a number) can be found, the function is said to be **differentiable**.

DERIVATIVE OF A GEOMETRIC FORMULA

The volume of a sphere with radius r, is given by: $V = \frac{4}{3}\pi r^3$

Find $V'(r)$.

❶ Set up the limit process.

$$V'(r) = \lim_{h \to 0} \frac{V(r+h) - V(r)}{h}$$

❷ Find $V(r+h)$ and $V(r)$ using $V = \frac{4}{3}\pi r^3$

$$= \lim_{h \to 0} \frac{\frac{4}{3}\pi(r+h)^3 - \frac{4}{3}\pi r^3}{h}$$

❸ Factor out the $\frac{4}{3}\pi$, expand the numerator, and then simplify.

$$= \frac{4}{3}\pi \cdot \lim_{h \to 0} \frac{(r+h)^3 - r^3}{h}$$

$$= \frac{4}{3}\pi \cdot \lim_{h \to 0} \frac{r^3 + 3r^2h + 3rh^2 + h^3 - r^3}{h}$$

$$= \frac{4}{3}\pi \cdot \lim_{h \to 0} \frac{3r^2h + 3rh^2 + h^3}{h}$$

❹ Factor out h in the numerator, simplify and then use direct substitution, 0 for h.

$$= \frac{4}{3}\pi \cdot \lim_{h \to 0} \frac{\cancel{h}(3r^2 + 3rh + h^2)}{\cancel{h}}$$

$$= \frac{4}{3}\pi \cdot \lim_{h \to 0}(3r^2 + 3rh + h^2)$$

$$= \frac{4}{3}\pi\left(3r^2 + 3r \cdot 0 + (0)^2\right)$$

$$= \frac{4}{3}\pi \cdot 3r^2$$

$$= 4\pi r^2$$

Therefore, $V'(r) = 4\pi r^2$

Let's say you want to find an equation of the line tangent to the graph of $f(x) = x^3 - 6x^2 + 9x - 13$ at the point with x-coordinate 2.

To find the equation of a line, two things are needed: a slope and a point on the line.

❶ To find the slope of the tangent line, use $f'(x)$, found in the "Derivative of a Cubic Polynomial" example earlier in this chapter

$$f(x) = x^3 - 6x^2 + 9x - 13$$
$$f'(x) = 3x^2 - 12x + 9$$

❷ Substitute 2 for x, in $f'(x)$, the derivative.

$$f'(2) = 3 \cdot (2)^2 - 12 \cdot 2 + 9$$
$$f'(2) = 12 - 24 + 9$$
$$f'(2) = -3$$

-3 is the slope of the line tangent at $x = 2$

❸ Next find the y-coordinate of the point with x-coordinate 2. Substitute 2 for x in the original function $f(x)$.

The point is $(2, -11)$

$$f(x) = x^3 - 6x^2 + 9x - 13$$

$$f(2) = 8 - 24 + 18 - 13$$
$$f(2) = -11$$

❹ Last, find the equation of the line having slope -3 and containing point $(2, -11)$.

Therefore, $y = -3x - 5$ is the equation of the line tangent to the graph of $f(x) = x^3 - 6x^2 + 9x - 13$ at the point with x-coordinate 2.

$$y - y_1 = m(x - x_1)$$
$$y - (-11) = -3(x - 2)$$
$$y + 11 = -3x + 6$$
$$y = -3x - 5$$

Horizontal Tangents

For many problems in calculus, you need to locate the horizontal tangent to a curve. An example of such a problem is finding the maximum and minimum values or a function. The slope (that is, derivative) at the point of tangency will be zero.

Find Points on a Curve at Which Tangent Line is Horizontal

Find the coordinates of each point on the graph of $f(x) = x^3 - 6x^2 + 9x - 13$ at which the tangent line is horizontal.

❶ The slope of the tangent line is given by $f'(x)$.

$$f(x) = x^3 - 6x^2 + 9x - 13$$
$$f'(x) = 3x^2 - 12x + 9$$

❷ The slope of the horizontal tangent is 0. Set $f'(x) = 0$ and solve for x.

$$0 = 3x^2 - 12x + 9$$
$$0 = 3(x^2 - 4x + 3)$$
$$0 = 3(x - 1)(x - 3)$$
$$\text{so } x = 1 \text{ or } x = 3$$

These are the x-coordinates of the points at which the tangent line is horizontal.

❸ Find the y-coordinates for the points with x-coordinates 1 and 3.

$$f(x) = x^3 - 6x^2 + 9x - 13$$
$$f(1) = 1^3 - 6 \cdot 1^2 + 9 \cdot 1 - 13$$
$$f(1) = -9$$
$$f(3) = 3^3 - 6 \cdot 3^2 + 9 \cdot 3 - 13$$
$$f(3) = -13$$

The points on the graph of $f(x) = x^3 - 6x^2 + 9x - 13$ at which the tangent lines are horizontal are $(1, -9)$ and $(3, -13)$.

❹ At the right is the graph of $f(x) = x^3 - 6x^2 + 9x - 13$ with horizontal red lines at $(1, -9)$ and $(3, -13)$.

Note: In Chapter 8 you will study the larger topic of *relative extrema* — the maximum and minimum values of a function. In this example, -9 is a relative maximum for $f(x)$, while -13 is a relative minimum.

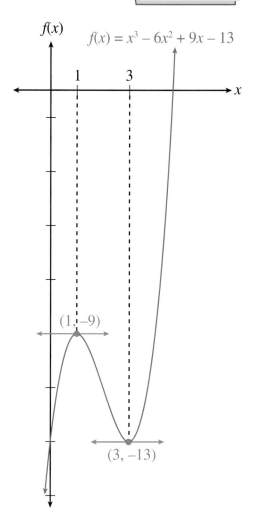

$f(x)$

$f(x) = x^3 - 6x^2 + 9x - 13$

1   3

x

$(1, -9)$

$(3, -13)$

Alternate Notations for a Derivative

There are many ways to indicate finding the derivative of the given function $y = f(x)$. Listed below in the left column are some directions you may encounter when doing a calculus problem. In the right column is the notation you would use as you write out your solution (as well a "pronunciation guide").

- Find the derivative of $f(x)$.

$f'(x)$ read "f prime of x"

- Find the derivative of $f(x)$.

$d/dx\, f(x)$ read "dee dee x of $f(x)$"

- For $y = f(x)$, find the derivative of y.

y' read "y prime"

- For $y = f(x)$ (that is, y is a function of x), find the derivative of y.

dy/dx read "dee y dee x" or "the derivative of y with respect to x"

Earlier in this section, for the function $f(x) = 3x^2 - 12x + 9$, it was determined that $f'(x) = 6x - 12$. You can write this fact in many ways.

For $f(x) = 3x^2 - 12x + 9$
$$f'(x) = 6x - 12$$
or
$$\frac{d}{dx} f(x) = 6x - 12$$

For $y = 3x^2 - 12x + 9$
$$y' = 6x - 12$$
or
$$\frac{dy}{dx} = 6x - 12$$

Some notations for the derivatives mentioned below are given in the right column.

- Second Derivative

$$y'', f''(x), \frac{d^2 y}{dx^2}, \frac{d^2}{dx^2}\left[f(x)\right]$$

- Third Derivative

$$y''', f'''(x), \frac{d^3 y}{dx^3}, \frac{d^3}{dx^3}\left[f(x)\right]$$

- nth Derivative

$$y^{(n)}, f^{(n)}(x), \frac{d^n y}{dx^n}, \frac{d^n}{dx^n}\left[f(x)\right]$$

$$\text{for } n \geq 4$$

Derivative as a Rate of Change

The derivative notation $\frac{dy}{dx}$ (or in other problems $\frac{dx}{dz}, \frac{dy}{dt}, \frac{dV}{dr}, \frac{dh}{dy}$) can also be interpreted as a rate of change. A few examples of derivative as a rate of change are air being pumped into a balloon, and the velocity and acceleration of a moving object.

BALLOON PROBLEM

Air is being pumped into a spherical balloon at the rate of 8π cubic inches per minute. Find the rate of change of the radius at the instant the radius is 2 inches.

❶ The information you are given is that $r = 2$ and $\frac{dV}{dt} = 8\pi$.

❷ Start with the formula for the volume of a sphere with radius r.

$$V(r) = \frac{4}{3}\pi r^3$$

❸ You determined $V'(r)$ in a previous example in this chapter, labeled as "Derivative of a Geometric Formula."

$$V'(r) = 4\pi r^2$$

❹ Use an alternate notation for $V'(r)$.

$$\frac{dV}{dr} = 4\pi r^2$$

❺ Multiply both sides of the equation by the term dr.

$$dV = 4\pi r^2 dr$$

❻ Divide both sides of the equation by the term dt (dt on the left and dt on the right).

$$\frac{dV}{dt} = 4\pi r^2 \frac{dr}{dt}$$

❼ Using the given information, substitute 8π for $\frac{dV}{dt}$ and 2 for r, and then solve the resulting equation for $\frac{dr}{dt}$.

When the radius is 2 inches, the radius is changing at the rate of $\frac{1}{2}$ inches per minute.

$$8\pi = 4\pi \cdot 2^2 \cdot \frac{dr}{dt}$$
$$8\pi = 16\pi \cdot \frac{dr}{dt}$$
$$\frac{8\pi}{16\pi} = \frac{dr}{dt}$$
$$\frac{1}{2} = \frac{dr}{dt}$$

Differentiability and Continuity

This chapter concludes with some comments on the relationship between differentiability and continuity.

When a Function Fails to Be Differentiable

The graph of a function can reveal points at which the function fails to be differentiable. This can occur at points at which the graph has a sharp turn, a vertical tangent, a "jump," or a "hole."

A GRAPH WITH A SHARP TURN

To the right is the graph of $f(x) = |1 - x^2|$, and a comment

about its differentiability at $x = \pm 1$.

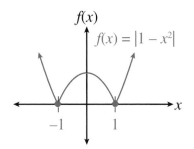

$f(x)$ is not differentiable at $x = -1$ and $x = 1$, since the slopes left and right of each of these numbers are not equal.

A GRAPH WITH A VERTICAL TANGENT LINE

To the right is the graph of $f(x) = x^{\frac{1}{3}}$, and a comment about its differentiability at $x = 0$.

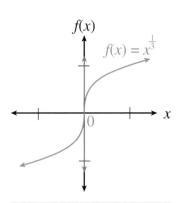

$f(x)$ is not differentiable at $x = 0$, because f has a vertical tangent at $x = 0$.

A GRAPH WITH A "JUMP"

To the right is the graph of $f(x) \begin{cases} 2 \text{ if } x > 2 \\ -1 \text{ if } x \leq 2 \end{cases}$ and a comment

about its differentiability at $x = 2$.

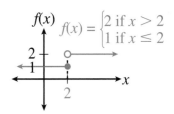

$$f(x) = \begin{cases} 2 \text{ if } x > 2 \\ 1 \text{ if } x \leq 2 \end{cases}$$

$f(x)$ is not differentiable at $x = 2$ because it is not continuous at $x = 2$ (there is a "jump" in the graph at $x = 2$).

A GRAPH WITH A "HOLE"

To the right is the graph of $f(x) = \dfrac{x^2 - 2x - 3}{x - 3}$ and a comment

about its differentiability at $x = 3$.

Note: The function at the right is not defined for $x = 3$, because it would result in a zero denominator.

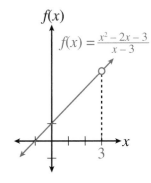

$$f(x) = \frac{x^2 - 2x - 3}{x - 3}$$

$f(x)$ is not differentiable at $x = 3$ because there is a "hole" in the graph.

Differentiability and Continuity *(continued)*

Relationship between Differentiability and Continuity

If the function f is differentiable at $x = c$ (that is in other words, $f'(c)$ exists there), then f is continuous at $x = c$.

Below are some graphs that we can analyze differentiability and continuity at a given point.

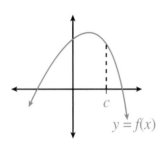

$f(x)$ is differentiable at $x = c$ and $f(x)$ is continuous at $x = c$.

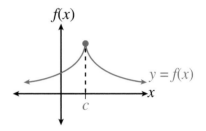

$f(x)$ is not differentiable at $x = c$, but $f(x)$ is continuous at $x = c$.

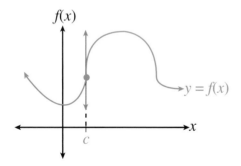

$f(x)$ is not differentiable at $x = c$, but $f(x)$ is continuous at $x = c$.

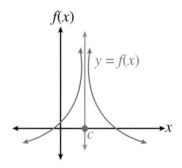

$f(x)$ is not differentiable at $x = c$, and $f(x)$ is not continuous at $x = c$.

chapter 4

Derivatives by Rule

In Chapter 3 you used the limit definition to find derivatives. In this chapter you will start to make use of many rules of differentiation, which enable you to find derivatives without using the time-consuming limit definition for derivatives.

Chapter 3 also introduces L'Hôpital's Rule — another technique to help you find limits that are one of the indeterminate forms.

Derivatives of Constant, Power, and Constant Multiple 78

Derivatives of Sum, Difference, Polynomial, and Product 80

The General Power Rule 84

The Quotient Rule 86

Rolle's Theorem and the Mean Value Theorem 89

Limits: Indeterminate Forms and L'Hôpital's Rule 93

Derivatives of Constant, Power, and Constant Multiple

This section introduces the Constant, Power, and Constant Multiple rules for finding derivatives and then includes an application of finding the equation of a line tangent to a curve.

The Constant, Power, and Constant Multiple Rules

THE CONSTANT RULE

If c is a constant, then $\frac{d}{dx}(c) = 0$.

$$\frac{d}{dx}(c) = 0$$

$$\frac{d}{dx}\left(5\pi\sqrt{3}\right) = 0$$

THE POWER RULE

If n is any rational number, then

$$\frac{d}{dx}(x^n) = n \cdot x^{n-1}.$$

$$\frac{d}{dx}(x^5) = 5x^4$$

If $y = \sqrt{x}$, then

$$y' = \frac{d}{dx}\left(\sqrt{x}\right)$$

$$= \frac{d}{dx}\left(x^{\frac{1}{2}}\right) = \frac{1}{2}x^{-\frac{1}{2}} = \frac{1}{2\sqrt{x}}$$

If $f(x) = \frac{1}{x^3}$, then

$$f'(x) = \frac{d}{dx}\left(\frac{1}{x^3}\right) = \frac{d}{dx}(x^{-3}) = -3x^{-4} = \frac{-3}{x^4}$$

THE CONSTANT MULTIPLE RULE

If c is a real number, and $f(x)$ is a differentiable function, then $\frac{d}{dx}\left[c \cdot f(x)\right] = c \cdot \left[\frac{d}{dx}(f(x))\right]$.

Just move the constant in front of the variable function. Next, multiply the constant by the function's derivative.

$$\frac{d}{dx}(5x^3) = 5 \cdot \frac{d}{dx}(x^3) = 5 \cdot 3x^2 = 15x^2$$

If $y = 4\sqrt{x}$, then

$$\frac{dy}{dx} = \frac{d}{dx}(4\sqrt{x}) = 4 \cdot \frac{d}{dx}(\sqrt{x})$$

$$= 4 \cdot \frac{d}{dx}\left(x^{-\frac{1}{2}}\right) = 4 \cdot \frac{1}{2} \cdot x^{-\frac{1}{2}} = \frac{2}{\sqrt{x}}$$

$$\frac{d}{dx}\left(\frac{6}{\sqrt[3]{x^2}}\right) = 6 \cdot \frac{d}{dx}\left(\frac{1}{\sqrt[3]{x^2}}\right) = 6 \cdot \frac{d}{dx}\left(\frac{1}{x^{\frac{2}{3}}}\right)$$

$$= 6 \cdot \frac{d}{dx}\left(x^{-\frac{2}{3}}\right) = 6 \cdot \frac{-2}{3} \cdot x^{-\frac{5}{3}}$$

$$= \frac{-4}{x^{\frac{5}{3}}} = \frac{-4}{\sqrt[3]{x^5}} \text{ or } \frac{-4}{\sqrt[3]{x^2}}$$

Derivatives of Sum, Difference, Polynomial, and Product

This section introduces the rules for finding the derivatives of sums and differences of functions, the derivative of a polynomial function, and the derivative of the product of two functions.

The Sum/Difference Rule

If f and g are both differentiable functions, then:

$$\frac{d}{dx}\left[f(x) \pm g(x)\right] = \frac{d}{dx}f(x) \pm \frac{d}{dx}g(x)$$

Note: A function is *differentiable* if its derivative can be found.

This can also be written as:

$$\frac{d}{dx}\left[f(x) \pm g(x)\right] = f'(x) \pm g'(x)$$

Or, using a popular *shorthand* notation, this could be written as:

$$\frac{d}{dx}\left(f \pm g\right) = f' \pm g'$$

$$\frac{d}{dx}\left(x^3 + 6x^2\right)$$

$$= \frac{d}{dx}\left(x^3\right) + \frac{d}{dx}\left(6x^2\right)$$

$$= 3x^2 + 12x$$

FIND THE DERIVATIVE OF A POLYNOMIAL FUNCTION

Using repeated applications of the first 4 differentiation rules—Constant, Constant Multiple, Sum/Difference, and Power—you can now find the derivative of any polynomial function.

If $f(x) = a_n x^n + a_{n-1}x^{n-1} + ... + a_2 x^2 + a_1 x + a_0$,
then $f'(x) = n \cdot a_n x^{n-1} + (n-1) \cdot$
$a_{n-1}x^{n-2} + ... + 2 \cdot a_2 x + a_1$

If $f(x) = 5x^3 - 6x^2 + 9x - 13$,

Then

$$f'(x) = 5 \cdot 3x^2 - 6 \cdot 2x + 9$$
$$= 15x^2 - 12x + 9$$

The Product Rule

If f and g are both differentiable functions, then:

$$\frac{d}{dx}\big[f(x) \cdot g(x)\big] = \overbrace{\frac{d}{dx}f(x)}^{\text{der. of 1st}} \cdot \overbrace{g(x)}^{\text{2nd}} + \overbrace{\frac{d}{dx}g(x)}^{\text{der. of 2nd}} \cdot \overbrace{f(x)}^{\text{1st}}$$

This can also be written as:

$$\frac{d}{dx}\big[f(x) \cdot g(x)\big] = f'(x) \cdot g(x) + g'(x) \cdot f(x)$$

Using shorthand notation, it can also be written as:

$$\frac{d}{dx}(f \cdot g) = f' \cdot g + g' \cdot f$$

If $h(x) = (3x^2 - 5x)(4x + 7),$

then $h'(x) = \overbrace{\left[\frac{d}{dx}(3x^2 - 5x)\right]}^{\text{der. of 1st}} \cdot \overbrace{(4x + 7)}^{\text{2nd}} + \overbrace{\left[\frac{d}{dx}(4x + 7)\right]}^{\text{der. of 2nd}} \cdot \overbrace{(3x^2 - 5x)}^{\text{1st}}$

$$= (6x - 5)(4x + 7) + 4(3x^2 - 5x)$$
$$= 24x^2 + 42x - 20x - 35 + 12x^2 - 20x$$
$$h'(x) = 36x^2 + 2x - 35$$

If you had first found the product of the two functions, you would have:

$h(x) = (3x^2 - 5x)(4x + 7)$
$h(x) = 12x^3 + x^2 - 35x,$
then find the derivative
$h'(x) = 36x^2 + 2x - 35$

Derivatives of Sum, Difference, Polynomial, and Product *(continued)*

ANOTHER PRODUCT RULE EXAMPLE

Find the derivative of $h(x) = \sqrt{x}(3x^2 - 6x)$.

① Start with the original function, rewriting $\sqrt{x}$ as a power.

$$h(x) = \sqrt{x}(3x^2 - 6x)$$
$$= x^{\frac{1}{2}}(3x^2 - 6x)$$

② Find h'(x) using the Product Rule.

$$h(x) = x^{\frac{1}{2}}(3x^2 - 6x)$$

$$h'(x) = \left[\overbrace{\frac{1}{2}x^{-\frac{1}{2}}}^{\text{der. of 1st}}\right] \cdot \overbrace{(3x^2 - 6x)}^{\text{2nd}} + \overbrace{(6x - 6)}^{\text{der. of 2nd}} \cdot \overbrace{\left(x^{\frac{1}{2}}\right)}^{\text{1st}}$$

③ Next, expand the terms on the right and simplify.

$$= \frac{3}{2}x^{\frac{3}{2}} - 3x^{\frac{1}{2}} + 6x^{\frac{3}{2}} - 6x^{\frac{1}{2}}$$
$$= \frac{15}{2}x^{\frac{3}{2}} - 9x^{\frac{1}{2}}$$

④ To finish up, factor $\frac{1}{2}x^{\frac{1}{2}}$ out of each term, and simplify to complete the process.

$$= \frac{1}{2}x^{\frac{1}{2}}(15x - 18)$$

$$h'(x) = \frac{\sqrt{x}}{2}(15x - 18)$$

ANOTHER LOOK AT THE PREVIOUS PROBLEM

Find the derivative of $h(x) = \sqrt{x}(3x^2 - 6x)$, without the Product Rule.

---

❶ Start with original function, rewriting $\sqrt{x}$ as a power; then distribute.

$$h(x) = \sqrt{x}(3x^2 - 6x)$$
$$= x^{\frac{1}{2}}(3x^2 - 6x)$$
$$h(x) = 3x^{\frac{5}{2}} - 6x^{\frac{3}{2}}$$

---

❷ Now find $h'(x)$ by finding the derivative of each term; then simplify.

$$h'(x) = 3 \cdot \frac{5}{2}x^{\frac{3}{2}} - 6 \cdot \frac{3}{2}x^{\frac{1}{2}}$$
$$= \frac{15}{2}x^{\frac{3}{2}} - \frac{18}{2}x^{\frac{3}{2}}$$

---

❸ As in the previous problem, factor $\frac{1}{2}x^{\frac{1}{2}}$ out of each

term, and simplify to complete the process.

$$= \frac{15}{2}x^{\frac{3}{2}} - \frac{18}{2}x^{\frac{1}{2}}$$
$$= \frac{1}{2}x^{\frac{1}{2}}(15x - 18)$$
$$= \frac{\sqrt{x}}{2}(15x - 18)$$

$$h'(x) = \frac{\sqrt{x}}{2}(15x - 18)$$

The General Power Rule

The General Power Rule gives you a means to find the derivative of the power of any function. It is a special case of the Chain Rule, which will be introduced in Chapter 5.

General Power Rule

If f is a differentiable function and n is a rational number, then:

$$\frac{d}{dx}\left[f(x)\right]^n = \overbrace{n \cdot \underbrace{\left(\text{inside function}\right)}_{}{}^{n-1}}^{} \cdot \overbrace{\frac{d}{dx} f(x)}^{\substack{\text{der. of inside} \\ \text{function}}}$$

This can also be written as:

$$\frac{d}{dx}\left[f(x)\right]^n = n \cdot \left[f(x)\right]^{n-1} \cdot f'(x)$$

Using shorthand notation, you can also write:

$$\frac{d}{dx}\left[f^n\right] = n \cdot f^{n-1} \cdot f'$$

$$\frac{d}{dx}\left(2x-3\right)^{10} = \overset{n}{10} \cdot \overbrace{\left(2x-3\right)^9}^{f^{n-1}} \cdot \overset{f'}{2}$$

$$= 20\left(2x-3\right)^9$$

It is usually helpful to write a radical expression as a power before attempting to compute its derivative.

If $f(x) = \sqrt[3]{4x+5}$, then

$$f'(x) = \frac{d}{dx}\sqrt[3]{4x+5}$$

$$= \frac{d}{dx}(4x+5)^{\frac{1}{3}}$$

$$= \underbrace{\frac{1}{3}}_{n} \overbrace{(4x+5)}^{(\text{inside})^{n-1}}{}^{-\frac{2}{3}} \cdot \overbrace{4}^{\substack{\text{der. of} \\ \text{inside}}}$$

$$= \frac{4}{3(4x+5)^{\frac{2}{3}}}$$

$$f'(x) = \frac{4}{3\sqrt[3]{4x+5}^2}$$

After writing the radical expression in the denominator as a power, move it up to the numerator to avoid having to use the General Power Rule.

$$\frac{d}{dx}\left(\frac{5}{\sqrt{x^2+3}}\right) = \frac{d}{dx}\frac{5}{(x^2+3)^{\frac{1}{2}}}$$

$$= \frac{d}{dx}\left[5 \cdot (x^2+3)^{-\frac{1}{2}}\right]$$

$$= 5 \cdot \frac{d}{dx}\left[(x^2+3)^{-\frac{1}{2}}\right]$$

$$= 5 \cdot \left[\underbrace{-\frac{1}{2}}_{n} \overbrace{(x^2+3)}^{(\text{inside})^{n-1}}{}^{-\frac{3}{2}} \cdot \overbrace{(2x)}^{\substack{\text{der. of} \\ \text{inside}}}\right]$$

The Quotient Rule

Like the Product Rule, the Quotient Rule involves putting pieces in the right places in the right formula and then simplifying the resulting expression. As in the Product Rule, your Algebra skills will be put to the test.

Statement of the Quotient Rule

If f and g are differentiable functions, and $g(x) \neq 0$, then:

$$\frac{d}{dx}\left(\frac{f(x)}{g(x)}\right) = \frac{\overbrace{\left[\frac{d}{dx}f(x)\right]}^{\text{der. of top}} \cdot \overbrace{g(x)}^{\text{bottom}} - \overbrace{\left[\frac{d}{dx}g(x)\right]}^{\text{der. of bottom}} \cdot \overbrace{g(x)}^{\text{top}}}{\underbrace{\left[g(x)\right]^2}_{\left(\text{bottom}\right)^2}}$$

This can also be written as:

$$\frac{d}{dx}\left(\frac{f(x)}{g(x)}\right) = \frac{f'(x) \cdot g(x) - g'(x) \cdot f(x)}{\left[g(x)\right]^2}$$

Or in shorthand notation, it can be written as:

$$\frac{d}{dx}\left(\frac{f}{g}\right) = \frac{f' \cdot g - g' \cdot f}{\left(g\right)^2}$$

If $h(x) = \dfrac{3x-2}{5x+4}$, then

$$h'(x) = \frac{\left[\frac{d}{dx}(3x-2)\right] \cdot (5x+4) - \left[\frac{d}{dx}(5x+4)\right] \cdot (3x-2)}{\left(5x+4\right)^2}$$

$$= \frac{3 \cdot (5x+4) - 5 \cdot (3x-2)}{\left(5x+4\right)^2}$$

$$= \frac{15x+12-15x+10}{\left(5x+4\right)^2}$$

$$h'(x) = \frac{22}{\left(5x+4\right)^2}$$

DERIVATIVE OF QUOTIENT OF RADICAL FUNCTIONS

Find $f'(x)$ for $f(x) = \dfrac{\sqrt{2x-5}}{\sqrt[3]{3x+1}}$.

❶ Rewrite each radical expression as a power.

$$f(x) = \frac{(2x-5)^{\frac{1}{2}}}{(3x+1)^{\frac{1}{3}}}$$

. . . .

❷ Find $f'(x)$ using the Quotient Rule.

$$\frac{d}{dx}\left(\frac{f}{g}\right) = \frac{f' \cdot g - g' \cdot f}{(g)^2}$$

$$f'(x) = \frac{\left[\dfrac{d}{dx}(2x-5)^{\frac{1}{2}}\right](3x+1)^{\frac{1}{3}} - \left[\dfrac{d}{dx}(3x+1)^{\frac{1}{3}}\right](2x-5)^{\frac{1}{2}}}{\left[(3x+1)^{\frac{1}{3}}\right]^2}$$

Note: *You found each derivative on the top by using the General Power Rule.*

$$= \frac{\dfrac{1}{2}(2x-5)^{-\frac{1}{2}} \cdot 2 \cdot (3x+1)^{\frac{1}{3}} - \dfrac{1}{3}(3x+1)^{-\frac{2}{3}} \cdot 3 \cdot (2x-5)^{\frac{1}{2}}}{(3x+1)^{\frac{2}{3}}}$$

$$= \frac{(2x-5)^{-\frac{1}{2}}(3x+1)^{\frac{1}{3}} - (3x+1)^{-\frac{2}{3}}(2x-5)^{\frac{1}{2}}}{(3x+1)^{\frac{2}{3}}}$$

❸ Next take out the common factor and then simplify the resulting expression.

$$= \frac{(2x-5)^{-\frac{1}{2}}(3x+1)^{-\frac{2}{3}}\left[(3x+1)-(2x-5)\right]}{(3x+1)^{\frac{2}{3}}}$$

$$= \frac{(2x-5)^{-\frac{1}{2}}(3x+1)^{-\frac{2}{3}}(x+6)}{(3x+1)^{\frac{2}{3}}}$$

$$= \frac{x+6}{(2x-5)^{\frac{1}{2}}(3x+1)^{\frac{4}{3}}}$$

❹ Last, rewrite the denominator in radical form.

$$f'(x) = \frac{x+6}{\sqrt{2x-5}\ \sqrt[3]{(3x+1)^4}}$$

$$f'(x) = \frac{x+6}{\sqrt{2x-5}(3x+1)\sqrt[3]{3x+1}}$$

This section covers two important theorems that relate continuity and differentiability: Rolle's Theorem and the Mean Value Theorem.

Rolle's Theorem

Let f be a function satisfying the following 3 conditions:

❶ f is continuous on the closed interval $[a,b]$

❷ f is differentiable on the open interval (a,b)

❸ $f(a) = f(b)$

Then, there exists at least one number c in (a,b) for which $f'(c) = 0$.

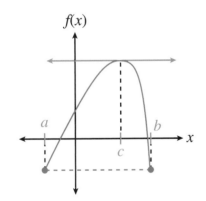

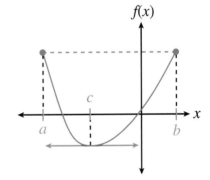

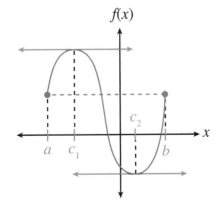

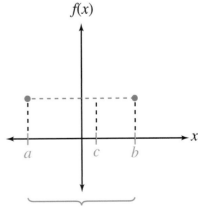

c is any number in (a, b)

For the function $f(x) = x^2 - 8x + 19$, find the value of c in the open interval $(2, 6)$ that is mentioned in Rolle's Theorem.

① To make use of Rolle's Theorem, you must first show that $f(x)$ satisfies all 3 conditions mentioned in the theorem:

Condition #1: Since $f(x) = x^2 - 8x + 19$ is a polynomial, it is everywhere continuous — so it is continuous on $[2, 6]$.

Condition #2: Since $f(x) = x^2 - 8x + 19$ is a polynomial, it is everywhere differentiable — so it is differentiable on $(2, 6)$.

Condition #3: After finding the values for $f(2)$ and $f(6)$, you can see that $f(2) = 7 = f(6)$.

② Since $f(x) = x^2 - 8x + 19$ satisfies the 3 conditions listed in the theorem, you can apply the conclusion:

There is at least one number c in $(2, 6)$ for which $f'(c) = 0$.

Find $f'(x)$.

$$f(x) = x^2 - 8x + 19$$
$$f'(x) = 2x - 8$$

③ Substitute c for x, set $f'(c) = 0$, and finish by solving for c.

$$f'(c) = 2c - 8$$
$$0 = 2c - 8$$
$$4 = c$$

4 is in the open interval $(2, 6)$ and $f'(4) = 0$.

④ At the right is the graph of $f(x) = x^2 - 8x + 19$, showing the horizontal tangent at $x = 4$ (where $f'(x) = 0$) in the interval $(2, 6)$.

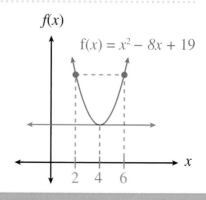

$f(x)$

$f(x) = x^2 - 8x + 19$

The Mean Value Theorem

Let f be a function which satisfies the following two conditions:

❶ f is continuous on the closed interval $[a,b]$

❷ f is differentiable on the open interval (a,b)

Then there exists at least one number c in (a,b) for which
$$f'(c) = \frac{f(b) - f(a)}{b - a}.$$

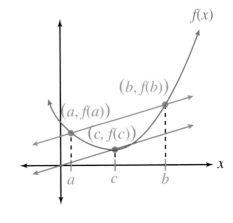

Note that the term $f'(c)$ is just the slope of the tangent to the graph of $f(x)$ at the point with coordinates $(c, f(c))$.

The other term on the right $\frac{f(b) - f(a)}{b - a}$ is the slope of the secant line containing the points $(a, f(a))$ and $(b, f(b))$.

For some number c in (a,b), the tangent line and the secant line have equal slope.

For $f(x) = x^3 - x^2 - 2x$, find the value of c in the interval $(-1,1)$, which is mentioned in the Mean Value Theorem.

❶ To make use of the Mean Value Theorem, you must first show that $f(x)$ satisfies both of the conditions mentioned in the theorem

Condition #1: Since $f(x) = x^3 - x^2 - 2x$ is a polynomial, it is everywhere continuous, so it is continuous on $(-1,1)$.

Condition #2: Since $f(x) = x^3 - x^2 - 2x$ is a polynomial, it is everywhere differentiable, so it is differentiable on $(-1,1)$.

Rolle's Theorem and the Mean Value Theorem *(continued)*

② Next, find $f'(x)$ and then replace the x with the c to get $f'(c)$.

$$f(x) = x^3 - x^2 - 2x$$
$$f'(x) = 3x^2 - 2x - 2$$
$$f'(c) = 3c^2 - 2c - 2$$

③ The interval in the problem is $(-1,1)$ so that $a = -1$ and $b = 1$.

Find the values of $f(a)$ and $f(b)$.

Using $f(x) = x^3 - x^2 - 2x$, you find that:

$$f(a) = f(-1)$$
$$= (-1)^3 - (-1)^2 - 2(-1)$$
$$= 0$$

and

$$f(b) = f(1)$$
$$= (1)^3 - (1)^2 - 2(1)$$
$$= -2$$

④ Now put all the pieces together and solve for c.

$$f'(c) = \frac{f(b) - f(a)}{b - a}$$
$$3c^2 - 2x - 2 = \frac{-2 - 0}{1 - (-1)}$$
$$3c^2 - 2x - 2 = -1$$
$$3x^2 - 2x - 1 = 0$$
$$(3c + 1)(c - 1) = 0$$

The $c = 1$ is *not* in the open interval $(-1,1)$.

Therefore $c = \frac{-1}{3}$ or $c = 1$

In Chapter 2 you encountered the indeterminate forms $\frac{0}{0}$ and $\frac{\infty}{\infty}$ when trying to calculate limits. These forms were dealt with by using tedious algebraic methods. L'Hôpital's Rule gives you a quicker alternative.

L'Hôpital's Rule

If $\lim\limits_{x \to c} \dfrac{f(x)}{g(x)}$ is one of the indeterminate forms $\frac{0}{0}$ and $\frac{\infty}{\infty}$, then $\lim\limits_{x \to c} \dfrac{f(x)}{g(x)} = \lim\limits_{x \to c} \dfrac{f'(x)}{g'(x)}$.

The indeterminate form $\frac{\infty}{\infty}$ may be one of the forms:

$$\frac{\infty}{\infty} \ or \ \frac{-\infty}{\infty} \ or \ \frac{\infty}{-\infty} \ \frac{-\infty}{-\infty}$$

TIP

Do not confuse this rule with the Quotient Rule. Here you are merely finding the derivative of the top and then the derivative of the bottom function and then finding the limit of their ratio.

L'HÔPITAL'S RULE: EXAMPLE 1

Determine $\lim\limits_{x \to 1} \dfrac{x^{12} - 1}{x^{11} - 1}$.

❶ Direct substitution leads to the indeterminate form $\frac{0}{0}$.

Apply L'Hôpital's Rule

$$\lim\limits_{x \to 1} \frac{x^{12} - 1}{x^{11} - 1} = \lim\limits_{x \to 1} \frac{12x^{11}}{11x^{10}}$$

❷ Now use direct substitution, 1 for x.

In Chapter 2, you would have divided all terms by x^{12}.

$$= \frac{12 \cdot 1^{11}}{11 \cdot 1^{10}}$$

$$= \frac{12}{11}$$

L'HÔPITAL'S RULE: EXAMPLE 2

Determine $\lim\limits_{x \to -3} \dfrac{\frac{1}{x} + \frac{1}{3}}{x + 3}$

❶ The indeterminate form $\dfrac{0}{0}$ results from direct substitution.

First rewrite the term $\dfrac{1}{x}$ as a power in preparation for finding the derivative of the top and the bottom.

$$\lim\limits_{x \to -3} \dfrac{\frac{1}{x} + \frac{1}{3}}{x + 3}$$

$$= \lim\limits_{x \to -3} \dfrac{x^{-1} + \frac{1}{3}}{x + 3}$$

❷ Apply L'Hôpital's Rule, taking the derivative of the top and the bottom.

$$= \lim\limits_{x \to -3} \dfrac{-1 \cdot x^{-2}}{1}$$

$$= \lim\limits_{x \to -3} \dfrac{-1}{x^{2}}$$

❸ Substitute $x = -3$.

In Chapter 2, you would have found a common denominator.

$$= \dfrac{-1}{9}$$

L'HÔPITAL'S RULE: EXAMPLE 3

Determine $\lim\limits_{x \to 4} \dfrac{2 - \sqrt{x}}{4 - x}$.

❶ After encountering the indeterminate form $\dfrac{0}{0}$, rewrite $\sqrt{x}$ as a power.

$$\lim\limits_{x \to 4} \dfrac{2 - \sqrt{x}}{4 - x}$$

$$= \lim\limits_{x \to 4} \dfrac{2 - x^{\frac{1}{2}}}{4 - x}$$

❷ Apply L'Hôpital's Rule – derivative of top and then derivative of bottom.

$$\lim_{x \to 4} \frac{\frac{-1}{2} x^{-\frac{1}{2}}}{-1}$$

$$= \lim_{x \to 4} \frac{1}{2\sqrt{x}}$$

❸ Substitute $x = 4$ and simplify.

In Chapter 2, you would have multiplied the numerator and denominator by the conjugate of the numerator.

$$= \frac{1}{2\sqrt{4}}$$

$$= \frac{1}{4}$$

L'HÔPITAL'S RULE: EXAMPLE 4

Calculate $\lim_{x \to 2} \dfrac{x^2 - x - 2}{x - 2}$

❶ Direct substitution leads to the indeterminate form $\dfrac{0}{0}$.

Use L'Hôpital's Rule.

$$\lim_{x \to 2} \frac{x^2 - x - 2}{x - 2}$$

$$= \lim_{x \to 2} \frac{2x - 1}{1}$$

$$= \lim_{x \to 2} (2x - 1)$$

❷ Put 2 in for x and then simplify.

In Chapter 2, you would have factored and reduced.

$$= 2 \cdot 2 - 1$$
$$= 3$$

FAQ

How do I know when to use L'Hôpital's Rule versus the techniques shown in Chapter 2?

If direct substitution leads to one of the indeterminates and you can find the derivative of both numerator and denominator, use L'Hôpital's Rule to find the limit.

chapter 5

Derivatives of Trigonometric Functions

In this chapter, you will greatly expand your ability to find derivatives—specifically derivatives of trigonometric and inverse trigonometric functions. L'Hôpital's Rule returns, and you are introduced to the **Chain Rule** (finding the derivative of a composite function).

Derivatives of Sine, Cosine, and Tangent 97

Derivatives of Secant, Cosecant, and Cotangent 100

L'Hôpital's Rule and Trigonometric Functions. 102

The Chain Rule 104

Trigonometric Derivatives and the Chain Rule 109

Derivatives of the Inverse Trigonometric Functions. 110

This section covers how to find the derivatives of three of the trigonometric functions: sine, cosine, and tangent. To the right is a "unit circle," which gives the cosine and sine of radian measures in the interval $[0, 2\pi]$.

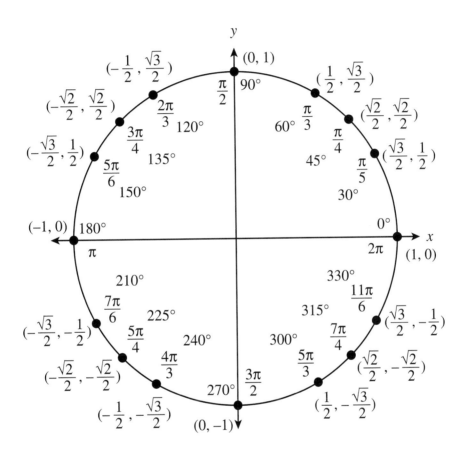

Sine and Cosine

Listed at right are the formulas for finding the derivatives of the sine and cosine functions. Following this text are some examples of how these two derivatives can be used.

$$\frac{d}{dx}(\sin x) = \cos x \quad \text{and} \quad \frac{d}{dx}(\cos x) = -\sin x$$

DERIVATIVE OF A SUM

Find $f'(x)$ for $f(x) = 3\sin x + 2\cos x$.

1 Start with the original function.

$$f(x) = 3\sin x + 2\cos x$$

2 Find the derivative using the Constant Multiple and Sum Rules (see Chapter 4):

$$f'(x) = 3 \cdot \frac{d}{dx}(\sin x) + 2 \cdot \frac{d}{dx}(\cos x)$$
$$= 3\cos x + 2(-\sin x)$$
$$f'(x) = 3\cos x - 2\sin x$$

DERIVATIVE OF A PRODUCT

Find $\frac{d}{dx}(\cos x \sin x)$.

1 Begin with the original expression.

$$\frac{d}{dx}(\cos x \sin x)$$

2 Apply the Product Rule.

$$= \left[\frac{d}{dx}\cos x\right]\sin x + \left[\frac{d}{dx}\sin x\right]\cos x$$

3 Find the derivatives of cosine and sine and then simplify.

$$= -\sin x \cdot \sin x + \cos x \cdot \cos x$$
$$= -\sin^2 x + \cos^2 x$$

DERIVATIVE OF PRODUCT OF ALGEBRAIC AND TRIGONOMETRIC FUNCTION

Find $f'(x)$ for $f(x) = x^2 \sin x$.

❶ Start with the original function.

$$f(x) = x^2 \sin x$$

❷ Apply the Product Rule (see Chapter 4).

$$f'(x) = (2x)\sin x + (\cos x) \cdot x^2$$

❸ Simplify.

$$= 2x\sin x + x^2 \cos x$$

DERIVATIVE OF A QUOTIENT

Find $f'(x)$ for $f(x) = \dfrac{1 + \cos x}{\sin x}$.

❶ Beginning with the original function, apply the Quotient Rule (see Chapter 4).

$$f(x) = \frac{1 + \cos x}{\sin x}$$

$$f'(x) = \frac{(-\sin x)\sin x - (\cos x)(1 + \cos x)}{(\sin x)^2}$$

❷ Simplify.

$$= \frac{-\sin^2 x - \cos x - \cos^2 x}{\sin^2 x}$$

❸ Group the $-\sin^2 x$ *and* $-\cos^2 x$ together and then use the Pythagorean Identity: $\sin^2 x + \cos^2 x = 1$.

$$= \frac{-\sin^2 x - \cos^2 x - \cos x}{\sin^2 x}$$

$$= \frac{-\left(\sin^2 x + \cos^2 x\right) - \cos x}{\sin^2 x}$$

$$= \frac{-1 - \cos x}{\sin^2 x}$$

Derivatives of Secant, Cosecant, and Cotangent

This section covers the derivatives of the other four trigonometric functions: tangent, secant, cosecant, and cotangent.

Formulas for Derivatives of Secant, Cosecant, and Cotangent

$$\frac{d}{dx} \tan x = \sec^2 x$$

$$\frac{d}{dx} \sec x = \sec x \tan x$$

$$\frac{d}{dx} \csc x = -\csc x \cot x$$

$$\frac{d}{dx} \cot x = -\csc^2 x$$

DERIVATIVE OF SECANT AT SPECIFIC VALUE

For $f(x) = \sec x$, find $f'\left(\frac{\pi}{3}\right)$.

1 Start with the original function and find its derivative.

$$f(x) = \sec x$$
$$f'(x) = \sec x \tan x$$

2 Find $f'\left(\frac{\pi}{3}\right)$.

$$f'\left(\frac{\pi}{3}\right) = \sec\left(\frac{\pi}{3}\right) \tan\left(\frac{\pi}{3}\right)$$

$$= \frac{1}{\cos\left(\frac{\pi}{3}\right)} \cdot \frac{\sin\left(\frac{\pi}{3}\right)}{\cos\left(\frac{\pi}{3}\right)}$$

$$= \frac{1}{\frac{1}{2}} \cdot \frac{\frac{\sqrt{3}}{2}}{\frac{1}{2}}$$

$$f'\left(\frac{\pi}{3}\right) = 2\sqrt{3}$$

DERIVATIVE OF ANOTHER PRODUCT

Find $f'(x)$ for $f(x) = \sec x \tan x$.

❶ Find the derivative using the Product Rule.

$$f(x) = \sec x \tan x$$

$$f'(x) = \left[\frac{d}{dx}\sec x\right]\tan x + \left[\frac{d}{dx}\tan x\right]\sec x$$

$$f(x) = \sec x \tan x + \sec^2 x \sec x$$

❷ Simplify and factor.

Note: *Although it is not required that you factor your derivatives, you will find in later chapters that it can be very helpful.*

$$= \sec x \tan^2 x + \sec^3 x$$

$$= \sec x(\tan^2 x + \sec^2 x)$$

DERIVATIVE OF A RADICAL TRIGONOMETRIC FUNCTION

Find y' for $y = \sqrt{2 + \tan x}$.

❶ Rewrite the original function as a power.

$$y = \sqrt{2 + \tan x}$$

$$= (2 + \tan x)^{\frac{1}{2}}$$

❷ Find y' using the General Power Rule.

$$y' = \frac{1}{2}(2 + \tan x)^{-\frac{1}{2}}\sec^2 x$$

❸ Simplify the result.

$$= \frac{\sec^2 x}{2(2 + \tan x)^{\frac{1}{2}}}$$

$$= \frac{\sec^2 x}{2\sqrt{2 + \tan x}}$$

L'Hôpital's Rule and Trigonometric Functions

This section covers L'Hôpital's Rule and its use in determining the limits of some of the trigonometric functions.

EXAMPLE 1

Determine $\lim\limits_{x \to 0} \frac{\sin x}{x}$.

Using direct substitution leads to the indeterminate form $\frac{0}{0}$.

❶ Apply L'Hôpital's Rule $\left(\lim \frac{f'}{g'} \right)$.

$$\lim_{x \to 0} \frac{\sin x}{x}$$

$$= \lim_{x \to 0} \frac{\cos x}{1}$$

❷ Use direct substitution, 0 for x.

$$= \frac{\cos(0)}{1}$$

$$= 1$$

EXAMPLE 2

Determine $\lim\limits_{x \to \frac{\pi}{2}} \dfrac{1 - \sin x}{\cos x}$.

Using direct substitution results in an indeterminate form.

❶ Apply L'Hôpital's Rule.

$$\lim_{x \to \frac{\pi}{2}} \frac{1 - \sin x}{\cos x}$$

$$= \lim_{x \to \frac{\pi}{2}} \frac{-\cos x}{-\sin x}$$

$$= \lim_{x \to \frac{\pi}{2}} \frac{\cos x}{\sin x}$$

❷ Use direct substitution, $\frac{\pi}{2}$ for x.

$$= \frac{\cos\left(\dfrac{\pi}{2}\right)}{\sin\left(\dfrac{\pi}{2}\right)}$$

$$= \frac{0}{1}$$

$$= 0$$

The Chain Rule

Frequently in your calculus studies, you will need to find the derivative of a composite function $f(g(x))$. This section discusses the Chain Rule—the tool to do this.

The Chain Rule: First Form

If f and g are both differentiable and H is the composite function defined by $H(x) = f(g(x))$, then H is differentiable and $H'(x)$ is given by:

$$H'(x) = \left[f'(g(x)) \right] \cdot g'(x)$$

Stated another way, you can write the Chain Rule as:

$$\frac{d}{dx} [f(g(x))] = [\; f'\; (g(x)) \; \cdot \; g'x)$$

der. of outer function, evaluated at inner funct. $=$ der. of outer function, evaluated at inner funct. times der. of inner function

In shorthand notation, it can be written as:

$$\text{If } y = f(g), \text{ then } y' = f'(g) \cdot g'$$

CHAIN RULE (FIRST FORM): EXAMPLE 1

Find $\frac{d}{dx}(\sin 3x)$.

❶ Begin with the original expression.

$$\frac{d}{dx}(\sin 3x)$$

❷ Identify the outer and the inner functions.

$$\frac{d}{dx}\left[\sin(3x)\right]$$

❸ Apply the Chain Rule and then simplify.

$$= \overbrace{\cos}^{\substack{\text{der. of} \\ \text{sin}}} \overbrace{(3x)}^{\text{at } 3x} \cdot \overbrace{3}^{\substack{\text{der. of} \\ 3x}}$$

$$= 3\cos 3x$$

CHAIN RULE (FIRST FORM): EXAMPLE 2

Find $\frac{d}{dx}(\cos x^3)$.

❶ Beginning with the original expression, identify the outer and inner functions.

$$\frac{d}{dx}(\cos x^3)$$

$$= \frac{d}{dx}(\cos x^3)$$

❷ Apply the Chain Rule and simplify.

$$= \overbrace{-\sin}^{\substack{\text{der. of} \\ \cos}} \overbrace{(x^3)}^{\text{at } x^3} \cdot \overbrace{3x^2}^{\substack{\text{der. of} \\ 3x^2}}$$

$$= -3x^2 \sin x^3$$

CHAIN RULE (FIRST FORM): EXAMPLE 3

Find $\frac{d}{dx}\sqrt{\sin x}$.

❶ Once again, identify the outer and inner functions.

$$\frac{d}{dx}\sqrt{\sin x}$$

$$= \frac{d}{dx}\sqrt{\sin x}$$

$$= \frac{d}{dx}(\sin x)^{\frac{1}{2}}$$

❷ Apply the Chain Rule and then simplify.

Note: You could also have just used the General Power Rule.

$$= \frac{1}{2}(\sin x)^{-\frac{1}{2}} \cdot \cos x$$

$$= \frac{\cos x}{2\sqrt{\sin x}}$$

....

CHAIN RULE (FIRST FORM): EXAMPLE 4

Calculate $\frac{d}{dx}\sin(\tan x)$.

❶ Rewrite the original expression, noting the outer and inner functions.

$$\frac{d}{dx}\sin(\tan x)$$

$$= \frac{d}{dx}\sin(\tan x)$$

....

❷ Apply the Chain Rule and simplify.

$$= \cos(\tan x) \cdot \sec^2 x$$

$$= \cos(\tan x)\sec^2 x$$

The Chain Rule: Second Form

If $y = f(u)$ is a differentiable function of u and if $u = g(x)$ is a differentiable function of x, then the composite function $y = f(g(x))$ is a differentiable of x, and $\frac{dy}{dx} = \frac{dy}{du} \cdot \frac{du}{dx}$.

Stated another way:

The derivative of y with respect to x equals the product of the derivative of y with respect to u and the derivative of u with respect to x.

CHAIN RULE (SECOND FORM): EXAMPLE 1

Find $\dfrac{dy}{dx}$ for $y = \sin 3x$.

❶ Start with the original function and write it as a composite of two functions.

$$y = \sin 3x$$
$$y = \sin(3x)$$

❷ Let y = the outer function and let u = the inner function.

$$y = \sin(u) \qquad u = 3x$$

❸ Find $\dfrac{dy}{du}$ and $\dfrac{du}{dx}$.

$$\frac{dy}{du} = \cos(u) \qquad \frac{du}{dx} = 3$$

❹ Apply the Chain Rule: Second Form.

$$\frac{dy}{dx} = \frac{dy}{du} \cdot \frac{du}{dx}$$
$$= \cos(u) \cdot 3$$
$$= 3\cos(u)$$

❺ Substitute $3x$ for u.

$$= 3\cos(3x)$$
$$\text{Therefore, } \frac{dy}{dx} = 3\cos 3x.$$

The Chain Rule
(continued)

CHAIN RULE (SECOND FORM): EXAMPLE 2

For $y = \sin^3 x$, find $\dfrac{dy}{dx}$.

① Rewrite the original function as a composite of two functions.

$$y = \sin^3 x$$
$$y = (\sin x)^3$$

② Let $y =$ the outer function and let $u =$ the inner function.

$$y = u^3 \qquad u = \sin x$$

③ Find $\dfrac{dy}{du}$ and $\dfrac{du}{dx}$.

$$\frac{dy}{du} = 3u^2 \qquad \frac{du}{dx} = \cos x$$

④ Apply the Chain Rule: Second Form.

$$\frac{dy}{dx} = \frac{dy}{du} \cdot \frac{du}{dx}$$
$$= 3u^2 \cdot \cos x$$
$$= 3u^2 \cdot \cos x$$

⑤ Substitute $\sin x$ for u.

$$= 3(\sin x)^2 \cos x$$
$$= 3\cos x \sin^2 x$$

Therefore, $\dfrac{dy}{dx} = 3\cos x \sin^2 x$.

After the use of the Chain Rule, the trigonometric derivatives, with u as a function of x, can now be written as follows.

$\Rightarrow \dfrac{d}{dx}(\sin u) = \cos u \; du$

$\dfrac{d}{dx}(\sin 5x) = \cos(5x) \cdot 5 = 5\cos x \, 5x$

$\Rightarrow \dfrac{d}{dx}(\cos u) = -\sin u \; du$

$\dfrac{d}{dx}\cos x^3 = -\sin x^3 \cdot 3x^2 = -3x^2 \sin x^3$

$\Rightarrow \dfrac{d}{dx}(\tan u) = \sec^2 u \; du$

$\dfrac{d}{dx}\tan \sqrt{x} = \sec^2 \sqrt{x} \cdot \dfrac{1}{2\sqrt{x}} = \dfrac{\sec^2 x}{2\sqrt{x}}$

$\Rightarrow \dfrac{d}{dx}(\csc u) = -\csc u \cot u \; du$

$\dfrac{d}{dx}\csc 3x = -\csc 3x \cot 3x \cdot 3 = -3\csc 3x \cot 3x$

$\Rightarrow \dfrac{d}{dx}(\sec u) = \sec u \tan u \; du$

$\dfrac{d}{dx}\sec x^2 = \sec x^2 \tan x^2 \cdot 2x = 2x \sec x^2 \tan x^2$

$\Rightarrow \dfrac{d}{dx}(\cot u) = -\csc^2 u \; du$

$\dfrac{d}{dx}\cot(x^4 + 7) = -\csc(x^4 + 7) \cdot 4x^3 \csc(x^4 + 7)$

Derivates of the Inverse Trigonometric Functions

Another set of derivatives you will need are those of the six inverse trigonometric functions.

INVERSE TRIGONOMETRIC FUNCTIONS

Read the last equation of "if $\sin y = u$, then $y = \arcsin u$," as "arc sine of u" or "inverse sine of u" (sometimes written as $y = \sin^{-1}u$). For example, since $\cos\left(\dfrac{\pi}{3}\right) = \dfrac{1}{2}$, you can write $\dfrac{\pi}{3} = \arccos\left(\dfrac{1}{2}\right)$, or $\dfrac{\pi}{3} = \cos^{-1}\left(\dfrac{1}{2}\right)$.

INVERSE TRIGONOMETRIC DERIVATIVES

If u is a function of x, then the derivative forms are as follows.

$$\frac{d}{dx}(\arcsin u) = \frac{du}{\sqrt{1 - u^2}} \qquad\qquad \frac{d}{dx}(\arcsin x^2) = \frac{2x}{\sqrt{1 - x^4}}$$

$$u = x^2, \text{ so } du = 2x$$

$$\Rightarrow \frac{d}{dx}(\arccos u) = \frac{-du}{\sqrt{1 - u^2}} \qquad\qquad \frac{d}{dx}(\arccos 3x) = \frac{-3}{\sqrt{1 - 9x^2}}$$

$$u = 3x, \text{ so } du = 3$$

$$\Rightarrow \frac{d}{dx}(\arctan u) = \frac{du}{1 + u^2} \qquad \frac{d}{dx}\left(\arctan\sqrt{x}\right) = \frac{\dfrac{1}{2\sqrt{x}}}{\sqrt{1 + \left(\sqrt{x^2}\right)}} = \frac{1}{2\sqrt{x}(1 + x)} \text{ with } u = \sqrt{x} \text{ then, } du = \frac{1}{2\sqrt{x}}$$

$\Rightarrow \dfrac{d}{dx}(\operatorname{arc\,cot} u) = \dfrac{-\,du}{1+u^2}$

$\dfrac{d}{dx}(\operatorname{arc\,cot} x^3) = \dfrac{-\,3x^2}{1+x^6}$ with $u = x^3,\ du= 3x^2$

$\Rightarrow \dfrac{d}{dx}(\operatorname{arc\,sec} u) = \dfrac{du}{|u|\sqrt{u^2-1}}$

$\dfrac{d}{dx}(\operatorname{arc\,sec} 3x) = \dfrac{3}{|3x|\sqrt{(3x)^2-1}} = \dfrac{1}{|x|\sqrt{9x^2-1}}$ for $u = 3x,\ du= 3$

$\Rightarrow \dfrac{d}{dx}(\operatorname{arc\,csc} u) = \dfrac{-\,du}{|u|\sqrt{u^2-1}}$

$\dfrac{d}{dx}(\operatorname{arc\,csc} x^2) = \dfrac{-\,2x}{|x^2|\sqrt{x^4-1}} = \dfrac{-\,2x}{x^2\sqrt{x^4-1}}$ Let $u = x^2$ so that $du = 2x$

chapter 6

Derivatives of Logarithmic and Exponential Functions

The first theme of this chapter is that of differentiating logarithmic functions—natural logarithmic as well as other base logarithmic functions. L'Hôpital's Rule is visited again, and then the chapter concludes with the derivative of exponential functions.

Derivatives of Natural Logarithmic
  Functions. 113
Derivatives of Other Base
  Logarithmic Functions 119
Logarithms, Limits, and
  L'Hôpital's Rule 123
Derivatives of Exponential
  Functions. 125

Derivatives of Natural Logarithmic Functions

The natural logarithmic function, written as *lnx*, has as its base the number *e*. The number *e* is defined many ways, and its approximate value is 2.71828.

At right are two of the more common ways of defining the number *e*.

$$e = \lim_{n \to \infty} \left(1 + \frac{1}{n}\right)^n \text{ or } e = \lim_{x \to 0} (1 + x)^{1/x}$$

In each case, you end up with an expression:
$(1+ \text{ really small number})^{\text{really big power}}$.

Instead of writing $\log_e x$, you just write ln*x*.

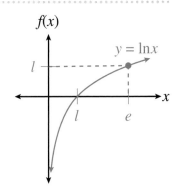

Properties of Natural Logarithms

Listed below are some properties of natural logarithms. These can be used to alter the form of a logarithmic expression or equation.

❶ If $\ln x = n$, then $e^n = x$. This shows the relationship between natural log equation and an exponential equation.

If $\ln x = 3$, then $x = e^3$.
If $x = e^{-2}$, then $\ln x = -2$.

❷ $\ln(xy) = \ln x + \ln y \Rightarrow$ the **log of a product property.**

$\ln(2x) = \ln 2 + \ln x$
$\ln(x^2) + \ln(y^3) = \ln(x^2 y^3)$

Derivatives of Natural Logarithmic Functions *(continued)*

❸ $\ln\left(\frac{x}{y}\right) = \ln x - \ln y \Rightarrow$ the log of a quotient property.=

the **log of a quotient property.**

$$\ln\left(\frac{2}{3}\right) = \ln 2 - \ln 3$$

$$\ln x - \ln 5 = \ln\left(\frac{x}{5}\right)$$

❹ $\ln x^n = n \cdot \ln x \Rightarrow$ the **log of a power property.**

$$\ln x^3 = 3\ln x$$

$$\ln\sqrt{x} = \ln x^{1/2} = \frac{1}{2}\ln x$$

$$\frac{2}{3}\ln x = \ln x^{2/3} = \ln\sqrt[3]{x^2}$$

❺ $\ln x = \dfrac{\log_b x}{\log_b e} \Rightarrow$ the **change of base property.**

$$\ln 5 = \frac{\log_{10} 5}{\log_{10} e}$$

$$\frac{\log_7 12}{\log_7 e} = \ln 12$$

Derivative of the Natural Logarithm Function

Listed below are the formulas used to find the derivative of $\ln x$, or $\ln u$ where u is some function of x. Following these formulas are some examples showing their uses in a variety of applications.

❶ $\dfrac{d}{dx}(\ln x) = \dfrac{1}{x}$

❷ If u is a differentiable function of x, then $\dfrac{d}{dx}(\ln u) = \dfrac{du}{u}$.

DERIVATIVE OF A NATURAL LOG OF A POWER

Find $f'(x)$ for $f(x) = \ln(x^2)$.

❶ This can be done one of two ways. Let's use derivative form number 2 (listed above) first.

Identify the u function.

$$f(x) = \ln(x^2) \text{ where } u = x^2$$
$$\text{and then } du = 2x$$

❷ Apply derivative form number 2 (see p. 114, "Derivative of the Natural Logarithm Function).

$$f'(x) = \frac{2x}{x^2} \Leftarrow \text{this is the } \frac{du}{u}$$

$$= \frac{2}{x}$$

....

A second way to approach the same problem is to take advantage of your log properties—specifically the log of a power property.

❶ Rewrite the original function using the log of a power property.

$$f(x) = \ln(x^2)$$

$$= 2 \cdot \ln x$$

....

❷ Use the natural log derivative form number 2 from above.

$$f'(x) = 2 \cdot \frac{1}{x}$$

$$= \frac{2}{x}$$

....

DERIVATIVE OF A LOG OF A RADICAL

Find $\frac{d}{dx}\left(\ln\sqrt{x+1}\right)$.

❶ Using the log of a power property, rewrite the original function.

$$\frac{d}{dx}\left(\ln\sqrt{x+1}\right)$$

$$= \frac{d}{dx}\ln(x+1)^{1/2}$$

$$= \frac{d}{dx}\left[\frac{1}{2} \cdot \ln(x+1)\right]$$

Letting $u = x + 1$, you have $du = 1$

....

❷ Find the derivative using the $\frac{du}{u}$ form.

$$= \frac{1}{2} \cdot \frac{1}{x+1}$$

$$= \frac{1}{2(x+1)}$$

Derivatives of Natural Logarithmic Functions *(continued)*

DERIVATIVE OF A QUOTIENT CONTAINING A NATURAL LOG

Find $f'(x)$ for $f(x) = \frac{\ln x}{x}$.

❶ Write the original function.

$$f(x) = \frac{\ln x}{x}$$

❷ Use the Quotient Rule to find the derivative.

$$f'(x) = \frac{\frac{1}{x} \cdot x - 1 \cdot \ln x}{(x)^2}$$

❸ Simplify the result.

$$f'(x) = \frac{1 - \ln x}{x^2}$$

DERIVATIVE OF A POWER OF A NATURAL LOG

Find $f'(x)$ for $f(x) = (\ln x)^3$.

❶ Start with a given function.

$$f(x) = (\ln x)^3$$

❷ Use the General Power Rule (or Chain Rule) to differentiate.

$$f'(x) = 3(\ln x)^2 \cdot \frac{1}{x}$$

$$f'(x) = \frac{3(\ln x)^2}{x}$$

DERIVATIVE OF A COMPLICATED NATURAL LOG EXPRESSION

Find $f'(x)$ for $f(x) = \ln\left[\dfrac{x(x^2+1)^2}{\sqrt{2x^3-1}}\right]$.

Here's where the real power of the natural log properties comes into use. You can write $\ln\left[\dfrac{x(x^2+1)^2}{\sqrt{2x^3-1}}\right]$ as a sum and/or a difference and/or a multiple of natural log expressions.

- -

① Begin with the function.

$$f(x) = \ln\left[\dfrac{x(x^2+1)^2}{\sqrt{2x^3-1}}\right]$$

- -

② Use the log of a quotient property.

$$= \ln\left[x(x^2+1)^2\right] - \ln\sqrt{2x^3-1}$$

- -

③ Use the log of a product property.

$$= \ln x + \ln(x^2+1)^2 - \ln\sqrt{2x^3-1}$$

- -

④ Rewrite the last term as a power.

$$= \ln x + \ln(x^2+1)^2 - \ln(2x-1)^{1/2}$$

- -

⑤ Use the log of a power property.

$$= \ln x + 2\ln(x^2+1) - \tfrac{1}{2}\ln(2x^3-1)$$

- -

⑥ Finally, find the derivative of each ln expression.
   Remember the $\frac{du}{u}$ for each ln derivative.

$$f'(x) = \tfrac{1}{x} + 2\left(\dfrac{2x}{x^2-1}\right) - \tfrac{1}{2}\left(\dfrac{6x^2}{2x^3-1}\right)$$

$$f'(x) = \tfrac{1}{x} + \dfrac{4x}{x^2-1} - \dfrac{3x^2}{2x^3-1}$$

Derivatives of Natural Logarithmic Functions *(continued)*

DERIVATIVE OF A TRIGONOMETRIC FUNCTION EVALUATED AT A NATURAL LOG

Find $f'(x)$ for $f(x) = \cos(\ln x)$.

❶ Start with the original function.

$$f(x) = \cos(\ln x) \quad \text{let } u = \ln x, \text{ then } du = \frac{1}{x}$$

❷ Find the derivative using $\frac{d}{dx}(\cos u) = -\sin u \cdot du$.

$$f'(x) = -\sin(\ln x) \cdot \frac{1}{x}$$
$$= \frac{-\sin(\ln x)}{x}$$

Derivatives of Other Base Logarithmic Functions

This section covers derivatives of logarithmic functions with bases other than e, the base of the natural logarithmic function. You write these logarithms as $\log_a x$ and read them as "the logarithm of x in base a" or as "the logarithm in base a of x."

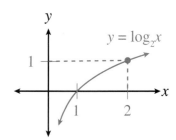

TIP

$\text{Log}_n 1 = 0$ for all positive bases n

$\text{Log}_7 1 = 0$ since $7^\circ = 1$

$\text{Log}_3 1 = 0$ since $3^\circ = 1$

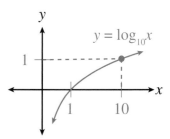

Properties of Logarithms

If x and y are positive numbers and $a > 0$, then the following properties of logarithms can be used. Notice the similarity of these properties to those of the natural logarithmic properties.

Property #1: If $\log_a x = n$, then $a^n = x$.

$\log_{10} 1,000 = 3$ because $10^3 = 1,000$

Since $\left(\dfrac{1}{2}\right)^4 = \dfrac{1}{16}$, then $\log_{1/2}\left(\dfrac{1}{16}\right) = 4$

If $\log_3\left(\dfrac{1}{9}\right) = -2$, then $3^{-2} = \dfrac{1}{9}$

Property #2: $\log_a(xy) = \log_a x + \log_a y$

$\log_{10}(99) = \log_{10}(9 \cdot 11) = \log_{10} 9 + \log_{10} 11 = \log 9 + \log 11$

Note: Logarithms in base 10 are called "common logs." Instead of writing $\log_{10} x$, you just write $\log x$.

$\log_2 3 + \log_2 5 = \log_2(3 \cdot 5) = \log_2 15$

Property #3: $\log_a\left(\dfrac{x}{y}\right) = \log_a x - \log_a y$

$$\log_3\left(\frac{10}{7}\right) = \log_3 10 - \log_3 7$$

$$\log_8 17 - \log_8 13 = \log_8\left(\frac{17}{13}\right)$$

Property #4: $\log_a x^n = n \cdot \log_a x$

$$\log_2 x^5 = 5 \cdot \log_2 x$$

$$\log_3\sqrt{x+1} = \log_3(x+1)^{1/2} = \frac{1}{2}\log_3(x+1)$$

$$\log_5\left(\frac{x^3 y^2}{\sqrt{z}}\right) = \log_5(x^3 y^2) - \log_5\sqrt{z}$$

$$= \log_5 x^3 + \log_5 y^2 - \log_5 z^{1/2}$$

$$= 3\log_5 x + 2\log_5 y - \frac{1}{2}\log_5 z$$

Property #5: $\log_a x = \dfrac{\log_b x}{\log_b a}$

$$\log_3 11 = \frac{\log_{10} 11}{\log_{10} 3} = \frac{\log 11}{\log 3}$$

$$\frac{\log_2 15}{\log_2 13} = \log_{13} 15$$

Derivatives of Logarithmic Functions

Listed below are the formulas for finding the derivative of just $\log_a x$, or $\log_a u$ where u is some function of *x*. Following these derivative formulas are some examples of how those formulas can be put to use.

❶ $\dfrac{d}{dx}(\log_a x) = \dfrac{1}{\ln a} \cdot \dfrac{1}{x}$

❷ If *u* is a differentiable function of *x*, then $\dfrac{d}{x}(\log_a u) = \dfrac{1}{\ln a}\dfrac{du}{u}$

DERIVATIVE OF A LOG OF A POLYNOMIAL

Find $f'(x)$ for $f(x) = \log_5(x^2 + 3)$.

① Start with the original function.

$$f(x) = \log (x^2 + 3)$$

② Identify the u and du.

Let $u = x^2 + 3$ and $du = 2x$

③ Find the derivative.

$$f'(x) = \frac{1}{\ln 5}\left(\frac{2x}{x^2 + 3}\right)$$

$$= \frac{2x}{(x^2 + 3)\ln 5}$$

DERIVATIVE OF LOG OF A QUOTIENT

Find $f'(x)$ for $f(x) = \log_3\left(\frac{x^3}{2x - 1}\right)$.

① Rewrite the original function using the log of a quotient property.

$$f(x) = \log_3\left(\frac{x^3}{2x - 1}\right)$$

$$= \log_3 x^3 - \log_3(2x - 1)$$

② Use the log of a power property on the first term.

$$= 3\log_3 x - \log_3(2x - 1)$$

❸ Find the derivative of each term using $\dfrac{1}{\ln a} \cdot \dfrac{du}{u}$.

$$f'(x) = 3 \cdot \dfrac{1}{\ln 3} \cdot \dfrac{1}{x} - \dfrac{1}{\ln 3} \cdot \dfrac{2}{2x - 1}$$

$$= \dfrac{3}{x \ln 3} - \dfrac{2}{(2x - 1) \ln 3}$$

$$= \dfrac{1}{\ln 3}\left(\dfrac{3}{x} - \dfrac{2}{2x - 1}\right)$$

DERIVATIVE OF A LOG OF A RADICAL FUNCTION

Find $f'(x)$ for $f(x) = \log_2 \sqrt[3]{x^3 - 5}$.

❶ Rewrite the radical as a power.

$$f(x) = \log_2 \sqrt[3]{x^3 - 5}$$

$$= \log_2(x^3 - 5)^{1/3}$$

❷ Use the log of a power property.

$$= \dfrac{1}{3}\log_2(x^3 - 5)$$

❸ Find the derivative.

$$f'(x) = \dfrac{1}{3} \cdot \dfrac{1}{\ln 2}\left(\dfrac{3x^2}{x^3 - 5}\right)$$

$$= \dfrac{x^2}{(x^3 - 5)\ln 2}$$

L'Hôpital's Rule returns here. You apply it to limits of natural log and a variety of other functions. It is restated below for your use:

$$\lim_{x \to c} \frac{f(x)}{g(x)} = \lim_{x \to c} \frac{f'(x)}{g'(x)} \text{ if the first limit is one of the indeterminate forms } \frac{0}{0} \text{ or } \frac{\infty}{\infty}.$$

LIMITS AND LOGS: EXAMPLE 1

Determine $\lim\limits_{x \to \infty} \dfrac{\ln(3x)}{x^2}$.

❶ Direct substitution leads the form $\dfrac{0}{0}$, so apply L'Hôpital's Rule.

$$\lim_{x \to \infty} \frac{\ln(3x)}{x^2}$$

$$= \lim_{x \to \infty} \frac{3/(3x)}{2x}$$

❷ Simplify and then use direct substitution.

$$= \lim_{x \to \infty} \frac{1}{2x^2}$$

$$= \frac{1}{\infty}$$

$$= 0$$

LIMITS AND LOGS: EXAMPLE 2

Determine $\lim\limits_{x \to \pi} \dfrac{\sin x}{\ln\left(\dfrac{x}{\pi}\right)}$.

❶ A $\dfrac{0}{0}$ form results when replacing x with π. L'Hôpital's Rule is applicable.

$$\lim\limits_{x \to \pi} \dfrac{\sin x}{\ln\left(\dfrac{x}{\pi}\right)}$$

$$= \lim\limits_{x \to \pi} \dfrac{\cos x}{\left(\dfrac{1}{\pi}\right) \div \left(\dfrac{x}{\pi}\right)}$$

❷ Simplify and substitute π for x.

$$= \lim\limits_{x \to \pi} x \cos x$$

$$= \pi \cos \pi$$

$$= \pi \cdot -1$$

$$= -\pi$$

Derivatives of Exponential Functions

This section introduces you to additional techniques of differentiating functions, such as $f(x) = 2\sqrt{x}$, $f(x) = e^{x^2}$, and $f(x) = 3^{\sin x}$. These are called **exponential functions**, with the base being a constant and an exponent that contains a variable.

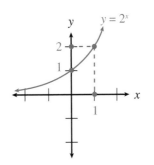

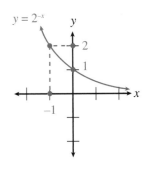

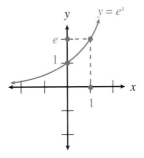

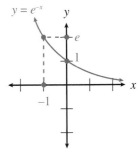

Rules for Differentiating Exponential Functions

There are two rules for differentiating exponential functions—one involves bases other than e, and the other involves e as the base.

Case I: The base is other than e.

If u is a differentiable function of x, then $\dfrac{d}{dx}(a^u) = a^u \cdot du \cdot \ln a$.

Using this formula, find $\dfrac{d}{dx}(3^{5x})$.

$$\underbrace{\frac{d}{dx}}_{der;\,of}\ \overbrace{3^{5x}}^{a^u} = \overbrace{3^{5x}}^{a^u} \cdot \overbrace{5}^{du} \cdot \overbrace{\ln 3}^{\ln a}$$

Derivatives of Exponential Functions *(continued)*

Case II: The base is e.

If u is a differentiable function of x, then $\dfrac{d}{dx}\left(e^u\right) = e^u \, du$.

Using this formula, find $\dfrac{d}{dx}\left(e^{x^3}\right)$.

$$\frac{d}{dx}\left(e^{x^3}\right) = e^{x^3} \cdot 3x^2$$

Finding the Derivative of Some Exponential Functions

Provided below are some additional examples that illustrate the use of the derivatives of exponential functions.

$y = 2^{\sin x}$

$y' = 2^{\sin x} \cdot \cos x \cdot \ln 2$

$f(x) = 3^{x^2 + 5x}$

$f'(x) = e^{x^2 + 5x} \cdot (2x + 5) \cdot \ln 3$

- - -

$f(x) = 7^{2x + 3}$

$f'(x) = 7^{2x + 3} \cdot 2 \cdot \ln 7$

   or

$f'(x) = 7^{2x + 3} \cdot \ln 7^2$

$\quad = 7^{2x + 3} \ln 49$

$y = 3^{\sqrt{\ln x}}$

$\quad = 3^{(\ln x)^{1/2}}$

$\quad = 3^{(1/2)\ln x}$

$y' = 3^{(1/2)\ln x} \cdot \left[\dfrac{1}{2}\left(\dfrac{1}{x}\right)\right] \cdot \ln 3$

$\quad = \dfrac{\ln 3 \cdot 3^{\sqrt{\ln x}}}{2x}$

- - -

$g(x) = x^3 e^{2x}$

$f'(x) = [3x^2] \cdot e^{2x} + [e^{2x} \cdot 2] \cdot x^3$

$\quad = 3x^2 e^{2x} + 2x^3 e^{2x}$

$\quad = x^2 e^{2x}(3 + 2x)$

$y = \dfrac{2^x}{e^x}$

$y' = \dfrac{(2^x \cdot 1 \cdot \ln 2) \cdot e^x - \left(e^x\right) \cdot 2^x}{\left(e^x\right)^2}$

$\quad = \dfrac{2^x e^x \ln 2 - e^x 2^x}{\left(e^x\right)^2}$

$\quad = \dfrac{2^x \ln 2 - 2^x}{e^x}$

Using L'Hôpital's Rule Revisited

We return to L'Hôpital's Rule and apply it to limits which involve exponential functions. L'Hôpital's Rule is restated below for your use:

$$\lim_{x \to c} \frac{f(x)}{g(x)} = \lim_{x \to c} \frac{f'(x)}{g'(x)}, \text{ if the first limit is one of the indeterminate forms } \frac{0}{0} \text{ or } \frac{\infty}{\infty}.$$

MORE LIMITS AND L'HÔPITAL'S RULE: EXAMPLE 1

Find $\lim\limits_{x \to 0} \dfrac{2^x - 7^x}{x}$.

1 Substituting 0 for x results in the indeterminate form $\frac{0}{0}$. Apply L'Hôpital's Rule.

$$\lim_{x \to 0} \frac{2^x - 7^x}{x}$$

$$= \lim_{x \to 0} \frac{2^x \ln 2 - 7^x \ln 7}{1}$$

2 Simplify.

$$= \lim_{x \to 0} \left(2^x \ln 2 - 7^x \lim 7 \right)$$

3 Use direct substitution and simplify again.

$$= 2^0 \ln 2 - 7^0 \ln 7$$

$$= \ln 2 - \ln 7$$

$$= \ln \left(\frac{2}{7} \right)$$

Derivatives of Exponential Functions *(continued)*

MORE LIMITS AND L'HÔPITAL'S RULE: EXAMPLE 2

Determine $\lim\limits_{x \to \infty} \dfrac{e^{3x}}{x^3}$.

❶ The original limit results in a $\frac{\infty}{\infty}$ indeterminate form, so apply L'Hôpital's Rule.

$$\lim_{x \to \infty} \frac{e^{3x}}{x^3}$$

$$= \lim_{x \to \infty} \frac{e^{3x} \cdot 3}{3x^2}$$

❷ Simplifying leads to another $\frac{\infty}{\infty}$ form, so apply L'Hôpital's Rule again.

$$= \lim_{x \to \infty} \frac{e^{3x}}{x^2}$$

$$= \lim_{x \to \infty} \frac{e^{3x} \cdot 3}{2x}$$

$$= \lim_{x \to \infty} \frac{3e^{3x}}{2x}$$

❸ Simplifying leads to another $\frac{\infty}{\infty}$ form, so apply L'Hôpital's Rule one more time.

$$= \lim_{x \to \infty} \frac{3\left(e^{3x} \cdot 3\right)}{2}$$

$$= \lim_{x \to \infty} \frac{9e^{3x}}{2}$$

$$= \infty$$

Therefore, no limit exists.

chapter 7

Logarithmic and Implicit Differentiation

This chapter introduces a technique called **logarithmic differentiation**—finding the derivative of a function having both a variable base and a variable exponent. Chapter 7 also covers **implicit differentiation**—finding the derivative of an equation having 2 or more variables for which it may be difficult or impossible to express one of the variables in terms of the other variables.

Logarithmic Differentiation 130

Techniques of Implicit
    Differentiation 134

Applications of Implicit
    Differentiation 139

Logarithmic Differentiation

Up to this point, you have been able to determine $\frac{d}{dx}(x^3)$ and $\frac{d}{dx}(3^x)$ in which the variable is in either the base or the exponent, but not in both places. A function of the form $y = [f(x)]^{g(x)}$, such as $f(x) = x^{\sin x}$ or $y = (\cos x)^{3x}$, can be differentiated using logarithmic differentiation—taking the natural log on both sides and then differentiating both sides.

EXAMPLE 1

Find y' for $y = (3x)^{x^2}$.

❶ Start with the original equation.

$$y = (3x)^{x^2}$$

❷ Take the natural log of both sides.

$$\ln y = \ln(3x)^{x^2}$$

❸ Use the log of a power property on the right.

$$\ln y = x^2 \ln(3x)$$

❹ Differentiate both sides—ln on the left, product rule on the right.

$$\frac{y'}{y} = [2x] \cdot \ln(3x) + \left[\frac{3}{3x}\right] \cdot x^2$$

❺ Simplify.

$$\frac{y'}{y} = 2x \cdot \ln(3x) + x$$

6 Multiply both sides by *y*.

$$y' = y(2x\ln(3x) + x)$$

7 Substitute $(3x)^{x^2}$ *for y*.

$$y' = (3x)^{x^2}(2x\ln(3x) + x)$$
$$\text{or}$$
$$y' = (3x)^{x^2}(\ln(3x)^{2x} + x)$$

EXAMPLE 2

Find $\dfrac{d}{dx}(\sin x)^{\cos x}$.

1 Let $y = (\sin x)^{\cos x}$.

$$y = (\sin x)^{\cos x}$$

2 Take ln of both sides.

$$\ln y = \ln(\sin x)^{\cos x}$$

3 Use the log of a power property on the right.

$$\ln y = \cos x \cdot \ln(\sin x)$$

4 Take the derivative of both sides—ln on the left, Product Rule on the right.

$$\frac{y'}{y} = [-\sin x] \cdot \ln(\sin x) + \left[\frac{\cos x}{\sin x}\right] \cdot \cos x$$

Logarithmic Differentiation
(continued)

5 Simplify.

$$\frac{y'}{y} = -\sin x \cdot \ln(\sin x) + \cot x \cdot \cos x$$

6 Multiply both sides by y.

$$y' = y[-\sin x \cdot \ln(\sin x) + \cot x \cdot \cos x]$$

7 Substitute $(\sin x)\cos x$ *for* y.

$$y' = (\sin x)^{\cos x}[-\sin x \cdot \ln(\sin x) + \cot x \cdot \cos x]$$
$$y' = (\sin x)^{\cos x}[\cot x \cdot \cos x - \sin x \cdot \ln(\sin x)]$$
$$y' = y' = (\sin x)^{\cos x}[\cot x \cdot \cos x - \ln(\sin x)^{\sin x}]$$

EXAMPLE 3

Find $f'(x)$ for $f(x) = (\ln x)^x$.

1 Starting with the original function, take the natural log of both sides.

$$f(x) = (\ln x)^x$$
$$\ln f(x) = \ln(\ln x)^x$$

2 Use the log of a product property and then differentiate the result.

$$\ln f(x) = x \cdot \ln(\ln x)$$

$$\frac{f'(x)}{f(x)} = [1] \cdot \ln(\ln x) + \left[\frac{\frac{1}{x}}{\ln x}\right] \cdot x$$

❸ Simplify and then multiply both sides by f(x).

$$\frac{f'(x)}{f(x)} = \ln(\ln x) + \frac{1}{\ln x}$$

$$f'(x) = f(x)\left[\ln(\ln x) + \frac{1}{\ln x}\right]$$

❹ Substitute (ln x)ˣ *for f(x).*

$$f'(x) = (\ln x)^x\left[\ln(\ln x) + \frac{1}{\ln x}\right]$$

FAQ

How do you know when to use "logarithmic differentiation"?

Use logarithmic differentiation when you are finding the derivative of a function such as $[f(x)]^{9(x)}$ and discover that both the base and exponent contain variables.

Techniques of Implicit Differentiation

Up to this point, the functions you encountered were expressed in an explicit form—that is, writing one variable in terms of another: $y = x^2 + 3x$, $s(t) = t^2 - t^2 + 15t - 7$, or $V(r) = \frac{4}{3}\pi r^3$. Unfortunately, many relationships are not written explicitly and are only implied by a given equation: $x^2 + y^2 = 25$, $xy = 7$, or $x + xy + 2y^3 = 13$. These equations are written in implicit form. It may not be possible to change an implicit form into an explicit form. For those cases, you use implicit differentiation.

Y AS A FUNCTION OF X ($\frac{dy}{dx}$ AS THE DERIVATIVE)

Find $\frac{dy}{dx}$ for $x^3 + xy - y^2 = 12$.

① Start with the original equation.

$$x^3 + xy - y^2 = 12$$

② Find the derivative of each term, treating y as a function of x.

$$\frac{d}{dx}(x^3) + \frac{d}{dx}(xy) - \frac{d}{dx}(y^2) = \frac{d}{dx}(12)$$

$$\underbrace{3x^2}_{\substack{power \\ rule}} + \overbrace{1 \cdot y + \frac{dy}{dx} \cdot x}^{product\ rule} - 2y\frac{dy}{dx} = 0$$

③ Isolate all $\frac{dy}{dx}$ terms of the left.

$$x\frac{dy}{dx} - 2y\frac{dy}{dx} = -3x^2 - y$$

④ Factor out $\frac{dy}{dx}$.

$$(x - 2y)\frac{dy}{dx} = -3x^2 - y$$

❺ Solve for $\dfrac{dy}{dx}$.

Note: We took to the opposite of both the top and the bottom, thus using fewer symbols in the final answer.

$$\frac{dy}{dx} = \frac{-3x^2 - y}{x - 2y}$$

or

$$\frac{dy}{dx} = \frac{3x^2 + y}{2y - x}$$

Y AS A FUNCTION OF *X* (*Y*′ AS THE DERIVATIVE)

Find y' for $(\ln x) \cdot y^3 = e^y \cdot x^2$.

❶ Start with the original equation.

$$(\ln x) \cdot y^3 = e^y \cdot x^2$$

❷ Differentiate implicitly, treating y as a function of x. Use the Product Rule on the left and on the right.

$$\left(\frac{1}{x}\right) \cdot y^3 + \left(3y^2 \cdot y'\right) \cdot \ln x = \left(e^y \cdot y'\right) \cdot x^2 + (2x) \cdot e^y$$

❸ Simplify.

$$\frac{y^3}{x} + 3y^2 \ln x \cdot y' = e^y x^2 \cdot y' + 2xe^y$$

❹ Isolate all y' terms on the left.

$$3y^2 \ln x \cdot y' - e^y x^2 \cdot y' = 2xe^y - \frac{y^3}{x}$$

❺ Factor out y' on the left.

$$\left(3y^2 \ln x - e^y x^2\right) y' = 2xe^y - \frac{y^3}{x}$$

❻ Solve for y'.

$$\left(3y^2 \ln x - e^y x^2\right) y' = 2xe^y - \frac{y^3}{x}$$

$$y' = \frac{xe^y - \dfrac{y^3}{x}}{3y^2 \ln x - e^y x^2}$$

. . . .

X AND *Y* AS FUNCTIONS OF AN UNKNOWN VARIABLE (*DX* AND *DY* AS THE DERIVATIVES)

Find $\dfrac{dy}{dx}$ for $e^{xy} + \sin x = \ln y$.

❶ Write the given equation.

$$e^{xy} + \sin x = \ln y$$

❷ Differentiate implicitly, treating x and y as functions of an unknown variable.

$$\overbrace{e^{xy}}^{e^u} \; \overbrace{\left(dx \cdot y + dy \cdot x\right)}^{du} + \cos x \cdot dx = \frac{dy}{y}$$

❸ Simplify.

$$e^{xy} ydx + e^{xy} xdy + \cos xdx = \frac{dy}{y}$$

❹ Put all dx terms on the left and all dy terms on the right.

$$e^{xy} ydx + e^{xy} xdy + \cos xdx = \frac{dy}{y}$$

$$e^{xy} ydx + \cos xdx = \frac{dy}{y} - e^{xy} xdy$$

5 Factor out dx on the left and dy on the right.

$$\left(e^{xy}y + \cos x\right)dx = \left(\frac{1}{y} - e^{xy}x\right)dy$$

6 Solve for $\frac{dy}{dx}$.

$$\left(e^{xy}y + \cos x\right)dx = \left(\frac{1}{y} - e^{xy}x\right)dy$$

$$\frac{e^{xy}y + \cos x}{\frac{1}{y} - e^{xy}x} = \frac{dy}{dx}$$

X AND *Y* AS FUNCTIONS OF *t* ($\frac{dx}{dt}$ AND $\frac{dy}{dt}$ AS THE DERIVATIVES)

Find $\frac{dy}{dx}$ for $-3xy + 2x^4 - 3y^3 = 5y$.

1 Begin with the original equation.

$$-3xy + 2x^4 - 3y^3 = 5y$$

2 Differentiate, treating both *x* and *y* as functions of some variable *t*.

$$\left(-3\frac{dx}{dt}\right)y + \left(\frac{dy}{dt}\right)(-3x) + 8x^3\frac{dx}{dt} - 9y^2\frac{dy}{dt} = 5\frac{dy}{dt}$$

3 Isolate all the $\frac{dy}{dt}$ terms on the left and all the $\frac{dx}{dt}$ terms on the right.

$$-3x\frac{dy}{dt} - 9y^2\frac{dy}{dt} - 5\frac{dy}{dt} = 3y\frac{dx}{dt} - 8x^3\frac{dx}{dt}$$

4 Factor out $\frac{dy}{dt}$ on the left and $\frac{dx}{dt}$ on the right.

$$\left(-3x - 9y^2 - 5\right)\frac{dy}{dt} = \left(3y - 8x^3\right)\frac{dx}{dt}$$

5 Multiply both sides by dt.

$$\left(-3x - 9y^2 - 5\right)\frac{dy}{dt} \cdot dt = \left(3y - 8x^3\right)\frac{dx}{dt} \cdot dt$$

6 Simplify.

$$(-3x - 9y^2 - 5)dy = (3y - 8x^3)dx$$

7 Solve for $\frac{dy}{dx}$.

$$\left(-3x - 9y^2 - 5\right)dy = \left(3y - 8x^3\,dx\right)$$

$$\frac{dy}{dx} = \frac{3y - 8x^3}{-3x - 9y^2 - 5}$$

or

$$\frac{dy}{dx} = \frac{8x^3 - 3y}{3x + 9y^2 + 5}$$

Applications of Implicit Differentiation

You can use the process of implicit differentiation in many word problems in calculus. Three examples follow—from differentiating a geometric formula, to solving a rate of change problem, to finding the equation of tangent line to a curve the equation of which is stated implicitly.

GEOMETRY FORMULA: V, R, AND H AS FUNCTIONS OF T ($\frac{dV}{dt}$, $\frac{dr}{dt}$, and $\frac{dh}{dt}$ AS THE DERIVATIVES)

Differentiate $V = \pi r^2 h$.

1 Start with the given formula.

$$V = \pi r^2 h$$

2 Differentiate implicitly, treating V, r, and h as functions of the variable t.

$$\frac{dV}{dt} = \left[2\pi r \frac{dr}{dt}\right] \cdot h + \left[\frac{dh}{dt}\right] \cdot \pi r^2$$

3 Simplify.

$$\frac{dV}{dt} = 2\pi rh \frac{dr}{dt} + \pi r^2 \frac{dh}{dt}$$

BALLOON RATE OF CHANGE PROBLEM USING IMPLICIT DIFFERENTIATION

A spherical balloon is being filled with air so that when its radius is 3 feet, the radius is increasing at the rate of $\frac{2}{3}$ ft./min. Find the rate of change of the volume at that instant.

1 Start with the formula for the volume, V, of a sphere in terms of its radius r.

$$V = \frac{4}{3}\pi r^3$$

2 Differentiate implicitly, treating both V and r as functions of time t.

$$\frac{dV}{dt} = \frac{4}{3}\pi 3 r^2 \frac{dr}{dt}$$

❸ Simplify.

$$\frac{dV}{dt} = 4\pi r^2 \frac{dr}{dt}$$

❹ Substitute $r = 3$ and $\frac{dr}{dt} = \frac{2}{3}$.

$$\frac{dV}{dt} = 4\pi (3)^2 \cdot \frac{2}{3}$$

❺ Simplify.

$$\frac{dV}{dt} = 24\pi \ ft^3 / \min$$

So, volume is increasing at a rate of $24\pi \text{ft}^3/\text{min}$.

LADDER SLIDING DOWN THE SIDE OF A BUILDING PROBLEM

A 20-foot ladder leans against the side of a building. The bottom of the ladder is 12 feet away from the bottom of the building and is being pulled away from the base of the building at a rate of 1.5 feet/second. Find the rate at which the distance from the top of the ladder to the base of the building is changing.

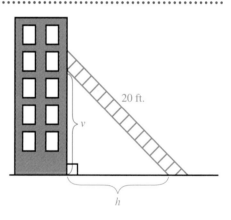

❶ Write a relationship between the vertical distance v and the horizontal distance h.

$$v^2 + h^2 = 20^2$$

❷ Differentiate implicitly, treating both v and h as functions of time, t.

$$2v \frac{dv}{dt} + 2h \frac{dh}{dt} = 0$$

❸ If it is known that $\frac{dh}{dt} = 1.5$ *and* $h = 12$, find a corresponding value of v by using the original equation (from the Pythagorean Theorem).

$$v^2 + h^2 = 20^2$$
$$v^2 + 12^2 = 20^2$$
$$v^2 + 144 = 400$$
$$v^2 = 256$$
$$v = 16$$

❹ Substitute values from Step 3 into the derivative from Step 2.

$$2v\frac{dv}{dt} + 2h\frac{dh}{dt} = 0$$
$$2(16)\frac{dv}{dt} + 2(12)(1.5) = 0$$

❺ Simplify and then solve for $\frac{dv}{dt}$.

$$32\frac{dv}{dt} + 36 = 0$$
$$\frac{dv}{dt} = -\frac{36}{32}$$
$$\frac{dv}{dt} = -\frac{9}{8}$$

The top of the ladder is sliding *down* (that's what the negative sign represents) the side of the building at a rate of $\frac{9}{8}$ ft./sec. But the answer to the question posed in this problem is that the distance from the top of the ladder to the base of the building is changing at a rate of $-\frac{9}{8}$ ft./sec.

chapter 8

Applications of Differentiation

This chapter shows you some of the many applications of a derivative. From horizontal tangents to equations of lines tangent to a curve, you will move on to critical numbers of a function and how to determine the intervals over which a function is increasing or decreasing. The chapter continues with finding extrema on a closed interval. From finding minimums and maximums over the complete domain of a function using the first derivative, you will move on to determining concavity, finding inflection points, and verifying relative extrema using the second derivative.

Tangent Line to Graph of a Function at a Point **143**

Horizontal Tangents **144**

Critical Numbers **146**

Increasing and Decreasing Functions **148**

Extrema of a Function on a Closed Interval **155**

Relative Extrema of a Function: First Derivative Test **160**

Concavity and Point of Inflection **165**

Extrema of a Function: Second Derivative Test **172**

You saw many examples of this in Chapter 3. Following is one new example just as a gentle reminder of the process used.

Tangent Line to the Graph of a Trigonometric Function

Find the equation of the line tangent to the graph of $f(x) = x^3 \ln x$ at the point with x coordinate e.

❶ Find $f(e)$.

$$f(x) = x^3 \ln x$$
$$f(e) = e^3 \ln e$$
$$f(e) = e^3 \cdot 1$$
$$f(e) = e^3 \text{ point is } (e, e^3)$$

❷ Find the slope at $x = e$.

$$f(x) = x^3 \ln x$$
$$f'(x) = \left[3x^2\right] \cdot \ln x + \left[\frac{1}{x}\right] \cdot x^3$$
$$f'(x) = 3x^2 \ln x + x^2$$
$$f'(e) = 3e^2 \ln e + e^2$$
$$= 3e^2 \cdot 1 + e^2$$
$$f'(e) = 4e^2 \text{ slope is } 4e^2$$

❸ Write the equation of the tangent line.

$$y - e^3 = 4e^2(x - e)$$
$$y - e^3 = 4e^2 x - 4e^3$$
$$y = 4e^2 x - 3e$$

Horizontal Tangents

A tangent line is horizontal when its slope is zero. After finding the derivative of a function, you set it equal to zero and then solve for the variable.

Horizontal Tangent to Graph of Polynomial Function

Find the coordinates of each point on the graph of $f(x) = x^3 - 12x^2 + 45x - 55$ at which the tangent line is horizontal.

1 Find $f'(x)$.

$$f(x) = x^3 - 12x^2 + 45x - 55$$
$$f'(x) = 3x^2 - 24x + 45$$

2 Set $f'(x) = 0$ and solve for x.

$$f'(x) = 3x^2 - 24x + 45$$
$$0 = 3x^2 - 24x + 45$$
$$0 = 3(x^2 - 8x + 15)$$
$$0 = 3(x - 3)(x - 5)$$
$$x = 3 \text{ or } x = 5$$

3 Using $f(x)$, find the corresponding y coordinate for each x in Step 2.

$$f(3) = -1, f(5) = -5$$

Tangent lines are horizontal at $(3, -1)$ and $(5, -5)$.

Horizontal Tangent to Graph of a Trigonometric Function

Find the x coordinate of each point on the graph of $f(x) = \cos 2x + 2\cos x$ in the interval $[x, 2\pi]$ at which the tangent line is horizontal.

1 Find $f'(x)$.

$$f(x) = \cos 2x + 2\cos x$$
$$f'(x) = -\sin(2x) \cdot 2 + 2(-\sin x)$$
$$f'(x) = -2\sin 2x - 2\sin x$$

2 Set $f'(x) = 0$.

$$0 = -2\sin 2x - 2\sin x$$

3 Replace $\sin 2x$ with $2\sin x \cos x$ and then simplify.

$$0 = -2(2\sin x \cos x) - 2\sin x$$
$$0 = -4\sin x \cos x - 2\sin x$$

4 Factor and then solve for x.

$$0 = -2\sin x(2\cos x + 1)$$
$$0 = -2\sin x \qquad 0 = 2\cos x + 1$$
$$0 = \sin x \qquad -\frac{1}{2} = \cos x$$

$$x = 0, x = \pi \qquad x = \frac{2\pi}{3}, x = \frac{4\pi}{3}$$

Critical Numbers

The critical numbers of a function play an important role in this chapter. You will use the critical numbers to help locate maximums and minimums for a function and inflection points for its graph.

DEFINITION OF CRITICAL NUMBERS

The number c is a **critical number** for $f(x)$ if and only if $f'(c) = 0$ or if $f'(c)$ is undefined.

In the first figure, the graph has a vertical asymptote at $x = 0$.

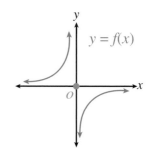

critical number at $x = 0$
$f'(o)$ undefined

In the second figure, a sharp corner occurs at $x = c$; the slopes of the curve left and right of $x = c$ are different — positive to the left, but negative to the right.

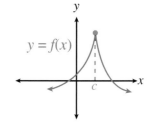

critical number at $x = c$
$f'(c)$ undefined

In the third figure, horizontal tangents to the graph occur at $x = a$ and $x = b$.

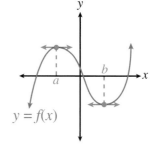

critical numbers at $x = a$ and $x = b$
$f'(a) = 0$
$f'(b) = 0$

CRITICAL NUMBERS OF A POLYNOMIAL FUNCTION

Find the critical numbers for $f(x) = 2x + 3x^2 - 6x + 4$.

❶ Find the derivative of the given function.

$$f(x) = 2x^3 + 3x^2 - 6x + 4$$
$$f'(x) = 6x^2 + 6x - 6$$

❷ Set $f'(x) = 0$. Other than taking out a common factor of 6, you can't factor further, so use the Quadratic Formula to solve for x.

$$0 = 6x^2 + 6x - 6$$
$$0 = 6(x^2 + x - 1)$$

$$x = \frac{-1 \pm \sqrt{5}}{2}$$

CRITICAL NUMBERS OF A RADICAL FUNCTION

Find the critical numbers for $f(x) = \sqrt{2x^2 - 8x}$.

❶ Rewrite the original function as a power.

$$f(x) = \sqrt{2x^2 - 8x}$$
$$f(x) = (2x^2 - 8x)^{1/2}$$

❷ Find $f'(x)$ using the General Power Rule (or Chain Rule).

$$f'(x) = \frac{1}{2}(2x^2 - 8x)^{-1/2}(4x - 8)$$
$$f'(x) = \frac{2x - 4}{\sqrt{2x^2 - 8x}}$$

❸ Set the numerator equal to 0; this is where $f'(x) = 0$. Set the denominator equal to 0; this is where $f'(x)$ is undefined.

$$0 = \frac{2x - 4}{\sqrt{2x^2 - 8x}}$$

$$0 = 2x - 4 \qquad 0 = \sqrt{2x^2 - 8x}$$

❹ Solve for x.

$$2 = x \qquad \qquad 0 = 2x^2 - 8x$$
$$0 = 2x(x - 4)$$
$$x = 0, \text{ or } x = 4$$

Increasing and Decreasing Functions

This section shows you how to make use of the function's derivative and critical numbers in order to find the intervals over which the functional values are increasing or decreasing.

DEFINITION OF INCREASING/DECREASING FUNCTION ON AN INTERVAL

❶ The function f is *increasing* on an open interval (a,b), if for any two numbers c and d in (a,b) with $c < d$, then $f(c) < f(d)$.

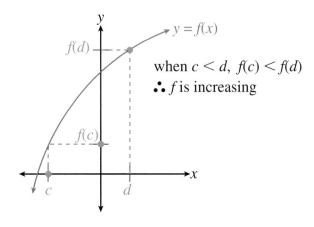

when $c < d,\ f(c) < f(d)$
∴ f is increasing

❷ The function f is *decreasing* on an open interval (a,b), if for any two numbers c and d in (a,b), with $c < d$, then $f(c) > f(d)$.

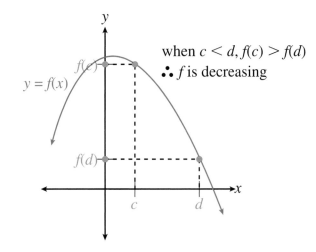

when $c < d,\ f(c) > f(d)$
∴ f is decreasing

PROPERTIES OF INCREASING/DECREASING FUNCTIONS ON AN INTERVAL

Let f be a continuous function on the closed interval $[a,b]$ and differentiable on the open interval (a,b).

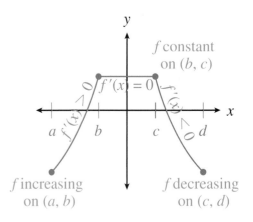

❶ If $f'(x) > 0$ for all x in (a,b), then f is increasing on $[a,b]$.

❷ If $f'(x) < 0$ for all x in (a,b), then f is decreasing on $[a,b]$.

❸ If $f'(x) = 0$ for all x in (a,b), the f is constant on $[a,b]$.

You now have a means to determine over which intervals a function is increasing, decreasing, or constant.

INCREASING/DECREASING INTERVALS FOR A POLYNOMIAL FUNCTION

Find the intervals over which $f(x)$ is increasing/decreasing for the function $f(x) = 2x^3 + 3x^2 - 12x$.

Note: *In the following examples, it is assumed that you will be able to use the appropriate methods to find the derivatives, which will merely be stated (but not derived step by step).*

❶ Find $f'(x)$ in simplified and factored form.

$$f(x) = 2x^3 + 3x^2 - 12x$$
$$f'(x) = 6x^2 + 6x - 12$$
$$f'(x) = 6(x^2 + x - 2)$$
$$f'(x) = 6(x + 2)(x - 1)$$

❷ Set $f'(x) = 0$ and solve for x.

$$0 = 6(x + 2)(x - 1)$$
$$x = -2, x = 1$$

❸ Using the zeros of $f'(x)$, create three open intervals and select a "test number" within each interval.

$x < -2, -2 < x < 1, and\ x > 1$

| $x = -3$ | $x = 0$ | $x = 2$ |

❹ Determine the sign of $f'(x)$ (+, –, or 0) at each "test number."

Note: *See Step 7 at the end of this problem. It shows how you can quickly determine the sign at each "test number" by using the factored form of $f'(x)$.*

$f'(x) = 6(x + 2)(x - 1)$		
$x < -2$	$-2 < x < 1$	$x > 1$
$f'(-3) = +-$	$f'(0) = ++$	$f'(2) = +++$
$f'(-3) > 0$	$f'(0) < 0$	$f'(2) > 0$
↗	↘	↗
inc.	dec.	inc.

❺ From the chart, you can easily identify the regions over which $f(x)$ is either increasing (↗) or decreasing (↘).

$f(x)$ is increasing for $x < -2$ and for $x > 1$ [or, in interval notation, $(-,-2)$ *and* $(1,)$].

$f(x)$ is decreasing for $-2 < x < 1$ [or, in interval notation, $(-2,1)$].

❻ To the right is the graph of $f(x) = 2x^3 + 3x^2 - 12x$.

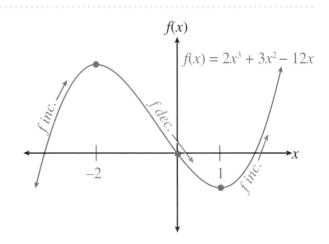

$f(x)$

$f(x) = 2x^3 + 3x^2 - 12x$

f inc.

f dec.

f inc.

-2

1

x

7 Here's the note about the sign determination of $f'(x)$ at each test number that was mentioned earlier:

Use the derivative $f'(x) = 6(x + 2)(x - 1)$ and the test numbers $x = -3$, $x = 0$, and $x = 2$.

FAQ

How do I know which "test numbers" to use in my computation?

Always use numbers that will make your computation easy.

1) If you want $x < e^{-1/3}$, use e^{-1}.

2) If you want $3 < x < \sqrt{17}$, use $x = 4$.

3) If you want $\dfrac{\pi}{4} < x < \dfrac{3\pi}{4}$, use $x = \dfrac{11}{2}$.

$$f'(x) = 6(x + 2)(x - 1)$$

$$f'(-3) = \overset{pos.}{\overbrace{6}}\,\overset{neg.}{\overbrace{(-3 + 2)}}\,\overset{neg.}{\overbrace{(-3 - 1)}} = \text{positive}$$

in shorthand notation, it looks like:

$$f'(-3) = + - - = +$$
$$f'(-3) > 0$$

so $f(x)$ is ↗ (inc.)

$$f'(x) = 6(x + 2)(x - 1)$$

$$f'(0) = \overset{pos.}{\overbrace{6}}\,\overset{pos.}{\overbrace{(0 + 2)}}\,\overset{neg.}{\overbrace{(0 - 1)}} = \text{negative}$$

in shorthand notation, it looks like:

$$f'(0) = + + - = -$$
$$f'(0) < 0$$

so $f(x)$ is ↘ (dec.)

$$f'(x) = 6(x + 2)(x - 1)$$

$$f'(2) = 6(2 + 2)(2 - 1) = + + + = \text{positive}$$

or in shorthand notation, it looks like:

$$f'(2) = + + +$$
$$f'(2) > 0$$

so $f(x)$ is ↗ (inc.)

....

INCREASING/DECREASING INTERVALS FOR A PRODUCT INVOLVING A NATURAL LOG FUNCTION

Find the intervals over which $f(x)$ is increasing/decreasing for the function $f(x) = x^3 \ln x$.

1 Find the simplified form of $f'(x)$.

$f(x) = x^3\ln x$ note that the domain of $f(x)$ is $x > 0$

$f'(x) = x^2(3\ln x + 1)$

. . . .

2 Set $f'(x) = 0$ and solve for x.

$0 = x^2(3\ln x + 1)$

$x^2 = 0$ $3\ln x + 1 = 0$

$x = 0$ $\ln x = -\dfrac{1}{3}$

$x = e^{-1/3}$

. . . .

3 Create the chart showing appropriate first derivative computations.

$f'(x) = x^2(3\ln x + 1)$	
$0 < x < e^{-1/3}$	$x > e^{-1/3}$
$f'(e^{-1}) = +-$	$f'(e) = ++$
$f'(e^{-1}) < 0$	$f'(e) > 0$
$\searrow$	$\nearrow$

. . . .

4 From the chart, you can easily identify the regions over which $f(x)$ is either increasing ($\nearrow$) or decreasing ($\searrow$).

. . . .

5 See the graph at right.

Note: $e^{-1/3} = \dfrac{1}{e^{1/3}} = \dfrac{1}{\sqrt[3]{e}}$

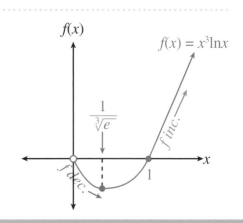

INCREASING/DECREASING INTERVALS FOR A PRODUCT OF TRIGONOMETRIC FUNCTIONS

Find the intervals over which $f(x)$ is increasing/decreasing for the function $f(x) = \sin x \cos x$ for $0 \le x \le 2\pi$.

❶ Find the simplified form of $f'(x)$.

$$f(x) = \sin x \cos x$$
$$f'(x) = \cos^2 x - \sin^2 x$$

❷ Set $f'(x) = 0$ and solve for x.

$$0 = \cos^2 x - \sin^2 x$$
$$\sin^2 x = \cos^2 x$$
$$\sin x = \pm \cos x$$
$$x = \frac{\pi}{4},\ x = \frac{3\pi}{4},\ x = \frac{5\pi}{4},\ x = \frac{7\pi}{4}$$

❸ Create the first derivative chart.

$f'(x) = \cos 2x - \sin 2x$				
$0 \le x < \frac{\pi}{4}$	$\frac{\pi}{4} < x < \frac{3\pi}{4}$	$\frac{3\pi}{4} < x < \frac{5\pi}{4}$	$\frac{5\pi}{4} < x < \frac{7\pi}{4}$	$\frac{7\pi}{4} < x < 2\pi$
$f'\left(\frac{\pi}{6}\right) > 0$	$f'\left(\frac{\pi}{2}\right) < 0$	$f'(\pi) > 0$	$f'\left(\frac{3\pi}{2}\right) < 0$	$f'\left(\frac{11\pi}{6}\right) > 0$
↗	↘	↗	↘	↗

❹ From the chart, you can easily identify the regions over which $f(x)$ is either increasing (↗) or decreasing (↘).

$f(x)$ is increasing for

$$0 \le x < \frac{\pi}{4}, \ \frac{3\pi}{4} < x < \frac{5\pi}{4}, \text{ and } \frac{7\pi}{4} < x < 2\pi .$$

$f(x)$ is decreasing for $\frac{\pi}{4} < x < \frac{3\pi}{4}$ and $\frac{5\pi}{4} < x < \frac{7\pi}{4}$.

❺ See the graph at right.

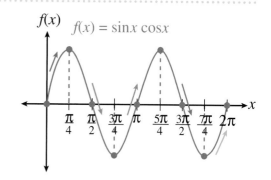

$f(x)$

$f(x) = \sin x \cos x$

The highs (maximums) and lows (minimums) of a function are known as its **extrema**. In this section, you learn how to locate these extrema on a closed interval, rather than on the entire domain of the function.

DEFINITION OF EXTREMA ON AN INTERVAL

If f is a function defined on an interval containing c, then:

1 $f(c)$ is a minimum of f on that interval, if $f(c) \leq f(x)$ for all x in that interval.

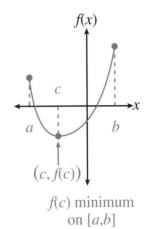

$f(c)$ minimum
on $[a,b]$

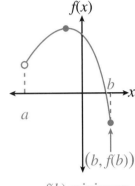

$f(b)$ minimum
on $[a,b]$

2 $f(c)$ is a maximum of f on that interval, if $f(c) \geq f(x)$ for all x in that interval.

maximum
on $[a,b]$

maximum
on $[a,b]$

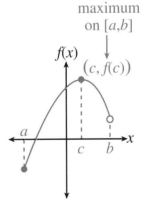

The extrema can occur at points within the interval or at an endpoint of an interval, called **endpoint extrema**.

FAQ

Another name for the minimum and maximum is **extrema**.

FINDING EXTREMA ON A CLOSED INTERVAL

Follow these steps to locate the extrema of continuous function f on a closed interval $[a,b]$.

❶ Find the critical numbers of f in $[a,b]$.

❷ Evaluate f at each critical number in $[a,b]$.

❸ Evaluate f at each endpoint of $[a,b]$.

❹ The smallest of these values is the minimum, and the largest of these values is the maximum.

EXTREMA OF A FUNCTION ON A CLOSED INTERVAL: POLYNOMIAL FUNCTION

Find the extrema of $f(x) = 5x^4 - 4x^3$ on the interval $[-1,2]$.

❶ Find $f'(x)$.

$$f(x) = 5x^4 - 4x^3$$
$$f'(x) = 20x^3 - 12x^2$$
$$f'(x) = 4x^2(5x - 3)$$

❷ Set $f'(x) = 0$ and solve for x—these are the critical numbers.

$$0 = 4x^2(5x - 3)$$
$$x = 0 \qquad x = \frac{3}{5}$$

❸ Evaluate $f(x)$ at each endpoint of the interval and at each critical number. (Note that you are *not* finding $f'(x)$ at each critical number.)

$f(x) = 5x^4 - 4x^3$			
left endpoint	critical number	critical number	right endpoint
$f(-1) = 9$	$f(0) = 0$	$f\left(\dfrac{3}{5}\right) = -\dfrac{27}{125}$	$f(2) = 48$
		minimum	maximum

❹ State the maximum and minimum values of $f(x)$ in the interval.

The maximum of f on $[-1,2]$ is 48 (at $x = 2$).

The minimum of f on $[-1,2]$ is $-\dfrac{27}{125}$ (at $x = \dfrac{3}{5}$).

❺ See the graph at right.

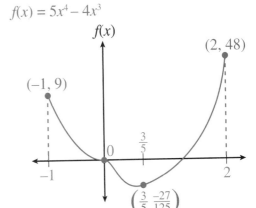

$f(x) = 5x^4 - 4x^3$

Extrema of a Function on a Closed Interval *(continued)*

EXTREMA OF A FUNCTION ON A CLOSED INTERVAL: TRIGONOMETRIC FUNCTION

Find the extrema of $f(x) = \sin^2 x + \cos x$ on the interval $[0, 2\pi]$.

❶ Find $f'(x)$.

$$f(x) = \sin^2 x + \cos x$$
$$= (\sin x)^2 + \cos x$$
$$f'(x) = 2(\sin x)\cos x - \sin x$$
$$f'(x) = 2\sin x \cos x - \sin x$$

❷ Set $f'(x) = 0$ and solve for x—these are the critical numbers.

$$0 = 2\sin x \cos x - \sin x$$
$$0 = \sin x (2\cos x - 1)$$

$$\sin x = 0 \qquad\qquad 2\cos x - 1 = 0$$

$$x = 0,\ x = \pi,\ x = 2\pi \qquad \cos x = \frac{1}{2}$$

$$x = \frac{\pi}{3} \qquad x = \frac{5\pi}{3}$$

❸ Evaluate $f(x)$ at each endpoint of the interval and at each critical number.

$f(x) = \sin^2 x + \cos x$				
left endpt. and crit. #	**critical #**	**crit. #**	**crit. #**	**right endpt. and crit. #**
$f(0) = 1$	$f\left(\dfrac{\pi}{3}\right) = \dfrac{5}{4}$	$f(\pi) = -1$	$f\left(\dfrac{5\pi}{3}\right) = \dfrac{5}{4}$	$f(2\pi) = 1$
	maximum	minimum	maximum	

④ State the maximum and minimum values of $f(x)$ in the interval.

The maximum of f on $[0,2\pi]$ is $\frac{5}{4}$ $\left(\text{at } \frac{\pi}{3} \text{ and } \frac{5\pi}{3}\right)$.

The minimum of f on $[0,2\pi]$ is -1 (at π).

⑤ See the graph at right.

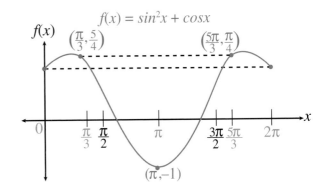

Relative Extrema of a Function: First Derivative Test

You are now ready to find the relative extrema over the entire domain of the function—not just on a closed interval, as in the last section.

DEFINITION OF RELATIVE EXTREMA

- $f(c)$ is called a **relative maximum** of f if there is an interval (a,b) containing c in which $f(c)$ is a maximum.

- $f(c)$ is called a **relative minimum** of f if there is an interval (a,b) containing c in which $f(c)$ is a minimum.

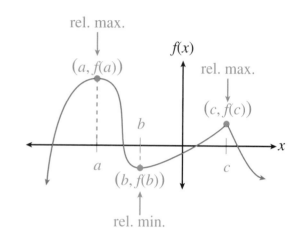

THE FIRST DERIVATIVE TEST

Let f be a function that is continuous on an open interval (a,b) containing a critical number c of f. If f is also differentiable on (a,b), except possibly at c, then:

- $f(c)$ is a relative minimum of f if $f'(x) < 0$ for $x < c$, but $f'(x) > 0$ for $x > c$.

- $f(c)$ is a relative maximum of f if $f'(x) > 0$ for $x < c$, but $f'(x) < 0$ for $x > c$.

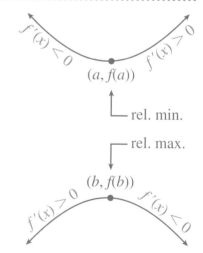

RELATIVE EXTREMA OF A POLYNOMIAL FUNCTION

Find the relative extrema of $f(x) = 3x^4 - 28x^3 + 60x^2$.

1 Find $f'(x)$.

$$f(x) = 3x^4 - 28x^3 + 60x^2$$
$$f'(x) = 12x^3 - 84x^2 + 120x$$
$$f'(x) = 12x(x^2 - 7x + 10)$$
$$f'(x) = 12x(x-2)(x-5)$$

2 Find the critical numbers of f by setting $f'(x) = 0$ and then solving for x.

$$0 = 12x(x-2)(x-5)$$
$$x = 0 \quad x = 2 \quad x = 5$$

3 Set up a first derivative chart to determine increasing or decreasing intervals for a function.

$f'(x) = 12x(x-2)(x-5)$			
$x < 0$	$0 < x < 2$	$2 < x < 5$	$x \geq 5$
$f'(-1) < 0$	$f'(1) > 0$	$f'(3) < 0$	$f'(6) > 0$
↘	↗	↘	↗
	rel. min.	rel. max.	rel. min.

4 Identify the x coordinates of the relative minimum/maximum of the function. Then find $f(x)$ for each of these x values.

The relative minimum occurs at $x = 0$ and at $x = 5$ ($f(0) = 0$ and $f(5) = -125$). So the relative minimum values are 0 and -125.

The relative maximum occurs at $x = 2$ ($f(2) = 64$). So the relative maximum value is 64.

5 The graph of $f(x) = 3x^4 - 28x^3 + 60x^2$ is shown at right.

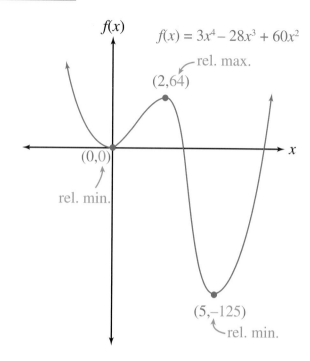

$f(x)$

$f(x) = 3x^4 - 28x^3 + 60x^2$

rel. max.

$(2,64)$

$(0,0)$

rel. min.

$(5,-125)$

rel. min.

x

RELATIVE EXTREMA OF A RATIONAL FUNCTION

Find the relative extrema of $f(x) = \dfrac{1 + x^2}{1 - x^2}.$

---

1 Find the derivative of $f(x)$.

$$f(x) = \frac{1 + x^2}{1 - x^2}$$

$$f'(x) = \frac{4x}{\left(1 - x^2\right)^2}$$

$x = \pm 1$ not in the domain of f

② Find the critical numbers for $f(x)$, setting both denominator and numerator equal to 0.

$$0 = \frac{4x}{\left(1 - x^2\right)^2}$$

$$4x = 0 \qquad \left(1 - x^2\right)^2 = 0$$

$$x = 0 \qquad 1 - x^2 = 0$$

$$x = \pm 1$$

③ Set up a first derivative chart to determine increasing and decreasing intervals.

$f'(x) = \dfrac{4x}{\left(1 - x^2\right)^2}$			
$x < -1$	$-1 < x < 0$	$0 < x < 1$	$x > 1$
$f'(-2) = \dfrac{-}{+}$	$f'\left(-\dfrac{1}{2}\right) = \dfrac{-}{+}$	$f'\left(\dfrac{1}{2}\right) = \dfrac{+}{+}$	$f'(2) = \dfrac{+}{+}$
$f'(-2) < 0$	$f'\left(-\dfrac{1}{2}\right) < 0$	$f'\left(\dfrac{1}{2}\right) > 0$	$f'(2) > 0$
↘	↘	↗	↗
		rel. min.	

④ Identify the x coordinates of the points at which the relative minimum/maximum occur.

The relative minimum occurs at $x = 0$ ($f(0) = 1$).

The relative maximum is 1.

5 To the right is the graph of
$$f(x) = \frac{1 + x^2}{1 - x^2}.$$

Note the vertical asymptotes at $x = -1$ and
$x = 1$. When $x = \pm 1$, the denominator of
$f(x)$ is zero.

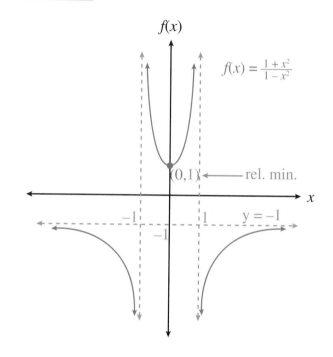

$f(x)$

$f(x) = \frac{1 + x^2}{1 - x^2}$

$(0,1)$ ← rel. min.

x

-1

1

$y = -1$

-1

The second derivative $(f''(x))$ allows you to find the intervals over which the graph of a function is concave up or concave down. The points at which the concavity changes (up to down, or down to up) locate points of inflection.

Definition of Concavity

- If the graph of *f* lies above all its tangents on an interval (a,b), then *f* is said to be **concave upward** on (a,b).

- If the graph of *f* lies below all its tangents on an interval (a,b), then *f* is said to be **concave downward** on (a,b).

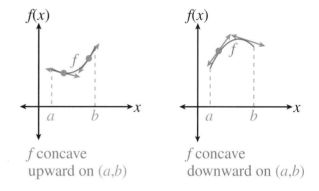

f concave upward on (a,b)

f concave downward on (a,b)

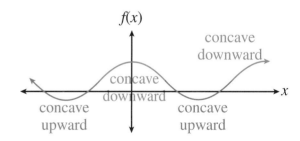

TEST FOR CONCAVITY

- If $f''(x) > 0$ for all *x* in (a,b), then the graph of *f* is concave upward on (a,b).

- If $f''(x) = < 0$ for all *x* in (a,b), then the graph of *f* is concave downward on (a,b).

Note: In the examples that follow, it is necessary to find both $f'(x)$ and $f''(x)$ for the given function. Since you have already seen many, many examples of finding derivatives, $f'(x)$ and $f''(x)$ will be merely stated—their derivations will not be shown here. Remember, $f''(x)$ is just the derivative of $f'(x)$.

Concavity and Point of Inflection *(continued)*

CONCAVITY FOR GRAPH OF A POLYNOMIAL FUNCTION

For the function $f(x) = x^4 + 2x^3 - 12x^2 - 15x + 22$, find the intervals over which its graph is concave upward or downward.

❶ Find $f'(x)$ and $f''(x)$.

$$f(x) = x^4 + 2x^3 - 12x^2 - 15x + 22$$
$$f'(x) = 4x^3 + 6x^2 - 24x - 15$$
$$f''(x) = 12x^2 + 12x - 24$$

❷ Set $f''(x) = 0$ and then solve for x.

$$0 = 12x^2 + 12x - 24$$
$$0 = 12(x^2 + x - 2)$$
$$0 = 12(x + 2)(x - 1)$$
$$x = -2, x = 1$$

❸ Create a second derivative chart using the critical numbers of $f'(x)$—that is, the zeros of $f''(x)$—to set up the appropriate intervals.

$f''(x) = 12(x + 2)(x - 1)$		
$x < -2$	$-2 < x < 1$	$x > 1$
$f''(-3) = + - -$	$f''(0) = + + -$	$f''(2) = + + +$
$f''(-3) > 0$	$f''(0) < 0$	$f''(2) > 0$
$\cup$	$\cap$	$\cup$
conc. up	conc. down	conc. up

❹ Identify the intervals of concavity:

The graph of f is concave upward when $x < -2$ and when $x > 1$.

The graph of f is concave downward when $-2 < x < 1$.

❺ To the right is the graph of
  $f(x) = x^4 + 2x^2 - 12x^2 - 15x + 22$.

Note: The graph will have two relative minimums and one relative maximum. Locate these points by using $f'(x)$ and a first derivative chart.

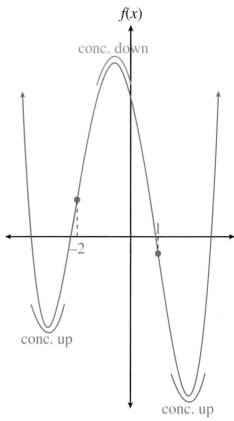

$f(x) = x^4 + 2x^3 - 12x^2 - 15x + 22$

$f(x)$

conc. down

conc. up

conc. up

Other features of concavity are listed below:

- The graph of *f* is concave upward on (*a*,*b*) if *f'* is increasing on (*a*,*b*).

Note: −, 0, + indicate slope of curve, i.e., $f^1(x)$

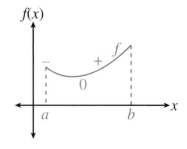

$f(x)$ increasing on (*a*,*b*)
<concave up>

- The graph of *f* is concave downward on (*a*,*b*) if *f'* is decreasing on (*a*,*b*).

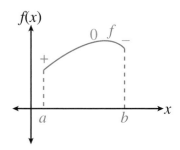

$f(x)$ decreasing on (*a*,*b*)
<concave down>

Definition of an Inflection Point

The point *P* is called a **point of inflection** for the graph of *f* if the concavity changes at the point *P*.

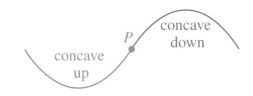

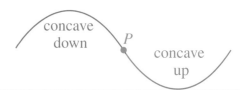

DETERMINING CONCAVITY AND FINDING AN INFLECTION POINT: POLYNOMIAL FUNCTION

Find the intervals over which the graph of $f(x) = 4x^3 - x^4$ is concave upward and downward, and find any inflection points.

① Find $f'(x)$ and $f''(x)$.

$$f(x) = 4x^3 - x^4$$
$$f''(x) = 12x^2 - 4x^3$$
$$f''(x) = 24x - 12x^2$$
$$f''(x) = 12x(2 - x)$$

② Set $f''(x) = 0$ and solve for x.

$$0 = 24x - 12x^2$$
$$0 = 12x(2 - x)$$
$$x = 0 \qquad x = 2$$

③ Create a second derivative chart.

$f''(x) = 12x(2-x)$		
$x < 0$	$0 < x < 2$	$x > 2$
$f''(-1) = -+$	$f''(1) = ++$	$f''(3) = +-$
$f''(-1) < 0$	$f''(1) > 0$	$f''(3) < 0$
$\cap$	$\cup$	$\cap$
conc. down	conc. up	conc. down

④ Identify the intervals of concavity and any points of inflection.

The graph of f is concave downward when $0 < x < 2$.
The graph of f is concave upward when $x < 0$ and when $x > 2$.
Since the concavity changes at $x = 0$ and then again at $x = 2$, these are the x coordinates of the points of inflection for the graph of f. $f(0) = 0$ and $f(2) = 16$.

Therefore, $(0,0)$ and $(2,16)$ are the points of inflection.

5 To the right is the graph of $f(x) = 4x^3 - x^4$.

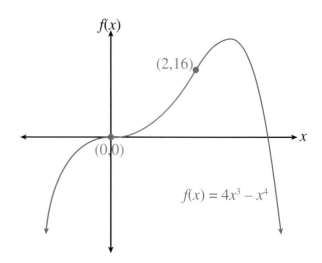

$f(x)$

$(2,16)$

$(0,0)$

x

$f(x) = 4x^3 - x^4$

DETERMINING CONCAVITY AND FINDING AN INFLECTION POINT: RADICAL FUNCTION

For the graph of $f(x) = x\sqrt{x+2}$, find the intervals of concavity and any inflection points.

1 Calculate $f'(x)$ and $f''(x)$.

$$f(x) = x\sqrt{x+2} \quad \text{domain of } f \text{ is } x > -2$$

$$f'(x) = \frac{3x+4}{2\sqrt{2x+2}}$$

$$f''(x) = \frac{3x+8}{4(x+2)^{3/2}}$$

2 Find the critical numbers of $f'(x)$ and set both the numerator and denominator of $f''(x) = 0$.

$$0 = \frac{3x+8}{4(x+2)^{3/2}}$$

$$3x + 8 = 0 \qquad 4(x+2)^{3/2} = 0$$

$$x = \frac{-8}{3} \qquad\qquad x = -2$$

❸ Prepare the second derivative chart. Notice that since the domain of f is $x > -2$, you will have only *one* column in your chart.

$$f''(x) = \frac{3x + 8}{4(x + 2)^{3/2}}$$

Note: Since $x = \frac{-8}{3}$ is not in the domain of the function, the only numbers to check are $x > -2$.

$$\underline{x > -2}$$

$$f''(0) = \frac{+}{+}$$

$$f''(0) > 0$$

$$\cup$$

conc. up

❹ Identify the intervals of concavity and any points of inflection.

The graph of f is concave upward for $x > -2$; in other words, *everywhere* in its domain. As such, there are *no* points of inflection.

❺ See the graph of $f(x) = x\sqrt{x + 2}$ at right.

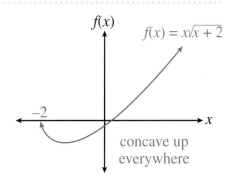

$f(x)$

$f(x) = x\sqrt{x + 2}$

-2

x

concave up everywhere

Extrema of a Function: Second Derivative Test

Sometimes you can avoid making a first derivative chart when trying to locate the relative minimum or maximum values of a function f. Using both the first and second derivatives allow you to save some time and work.

The Second Derivative Test for Relative Extrema

Let f be a function for which $f'(c) = 0$ and the second derivative of f exists at c. Then:

- If $f''(c) > 0$, then $f(c)$ is a relative minimum.

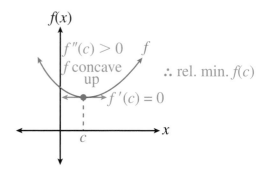

- If $f''(c) < 0$, then $f(c)$ is a relative maximum.

- If $f''(c) = 0$, then the second derivative test fails; you must use the first derivative test instead.

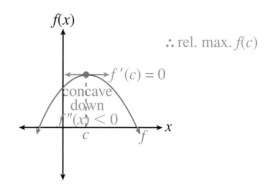

SECOND DERIVATIVE TEST AND RELATIVE EXTREMA OF A POLYNOMIAL FUNCTION

Find the relative extrema for $f(x) = x^3 - 3x^2 - 9x + 7$.

① Find $f'(x)$.

$$f(x) = x^3 - 3x^2 - 9x + 7$$
$$f'(x) = 3x^2 - 6x - 9$$

② Set $f'(x) = 0$ and then solve for x.

$$0 = 3x^2 - 6x - 9$$
$$0 = 3(x^2 - 2x - 3)$$
$$0 = 3(x + 1)(x - 3)$$
$$x = -1 \qquad x = 3$$

③ Find $f''(x)$.

$$f'(x) = 3x^2 - 6x - 9$$
$$f''(x) = 6x - 6$$

④ Find $f''(x)$ for each value of x in Step 2.

$$f''(-1) = 6(-1) - 6 \qquad f''(3) = 6(3) - 6$$
$$f''(-1) = -12 \qquad f''(3) = 12$$
$$f''(-1) < 0 \qquad f''(3) > 0$$

Since $f'(-1) = 0$ and $f''(-1) < 0$, there is a relative maximum at $x = -1$.
Since $f'(3) = 0$ and $f''(3) > 0$, there is a relative minimum at $x = 3$.

⑤ Find $f(-1)$ and $f(3)$.

$$f(x) = x^3 - 3x^2 - 9x + 7$$
$$f(-1) = 12 \qquad f(3) = -20$$

Therefore, 12 is the relative maximum and −20 is the relative minimum.

6 The graph of $f(x) = x^3 - 3x^2 - 9x + 7$ is shown at right.

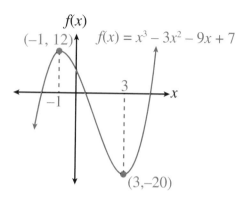

$f(x)$

$(-1, 12)$ $f(x) = x^3 - 3x^2 - 9x + 7$

3

-1

x

$(3, -20)$

SECOND DERIVATIVE TEST FOR EXTREMA: TRIGONOMETRIC FUNCTION

For $f(x) = \sin x + \cos x$, find the relative extrema on the interval $[0, 2\pi]$.

---

1 Find the first derivative.

$$f(x) = \sin x + \cos x$$
$$f'(x) = \cos x - \sin x$$

---

2 Set $f'(x) = 0$ and solve for x.

$$0 = \cos x - \sin x$$
$$\sin x = \cos x$$
$$x = \frac{\pi}{4} \text{ and } x = \frac{5\pi}{4}$$

---

3 Find the second derivative.

$$f'(x) = \cos x - \sin x$$
$$f''(x) = -\sin x - \cos x$$

4 Find $f''\left(\frac{\pi}{4}\right)$ and $f''\left(\frac{5\pi}{4}\right)$.

$$f''\left(\frac{\pi}{4}\right) = -\sin\frac{\pi}{4} - \cos\frac{\pi}{4} \qquad f''\left(\frac{5\pi}{4}\right) = -\sin\frac{5\pi}{4} - \cos\frac{5\pi}{4}$$

$$f''\left(\frac{\pi}{4}\right) = -\frac{\sqrt{2}}{2} - \frac{\sqrt{2}}{2} \qquad f''\left(\frac{5\pi}{4}\right) = -\left(-\frac{\sqrt{2}}{2}\right) - \left(-\frac{\sqrt{2}}{2}\right)$$

$$f''\left(\frac{\pi}{4}\right) = -\sqrt{2} \qquad f''\left(\frac{5\pi}{4}\right) = \sqrt{2}$$

$$f''\left(\frac{\pi}{4}\right) < 0 \qquad f''\left(\frac{5\pi}{4}\right) > 0$$

5 Find $f\left(\frac{\pi}{4}\right)$ and $f\left(\frac{5\pi}{4}\right)$.

$$f\left(\frac{\pi}{4}\right) = \sqrt{2} \qquad f\left(\frac{5\pi}{4}\right) = -\sqrt{2}$$

With $f'\left(\frac{\pi}{4}\right) = 0$ and $f''\left(\frac{\pi}{4}\right) < 0$, there is a relative maximum at $x = \frac{\pi}{4}$. Therefore, $\sqrt{2}$ is the relative maximum.

With $f'\left(\frac{5\pi}{4}\right) = 0$ and $f''\left(\frac{5\pi}{4}\right) > 0$, there is a relative minimum at $x = \frac{5\pi}{4}$. Therefore, $-\sqrt{2}$ is the relative minimum.

6 At right is the graph of $f(x) = \sin x + \cos x$.

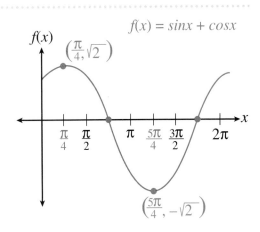

$$f(x) = sinx + cosx$$

chapter 9

Additional Applications of Differentiation: Word Problems

The derivative is used to solve many types of word problems in calculus. The first type of word problem covered in this chapter is optimization, in which you are asked to determine such things as the largest volume or the least cost. The second type of problem contains related rates in which you will find the rate at which the water level in a tank is changing or the rate at which the length of a shadow is changing. The last type of word problem requires you to use the derivative to go from a position function to its velocity and its acceleration functions.

Optimization. 177

Related Rates 183

Position, Velocity, and Acceleration . . . 188

A common application of the use of the derivative in calculus is determining the minimum and maximum values of a function which describes a word problem—for example, the largest area, least time, greatest profit, or the most optimal dimensions.

Volume of a Box Problem

Let's say you are cutting equal squares from each corner of a rectangular piece of aluminum that is 16 inches by 21 inches. You will then fold up the "flaps" to create a box with no top. Find the size of the square that must be cut from each corner in order to produce a box having maximum volume.

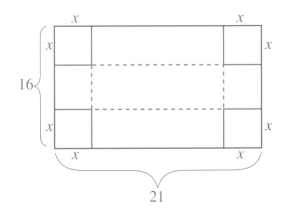

❶ Let x be the length of a side of each square to be removed. After the squares are removed from each corner, the aluminum now looks like the figure at right.

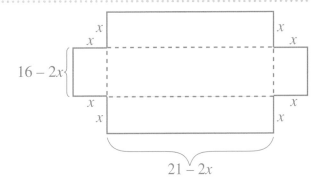

❷ When the flaps are folded up, the box has the dimensions shown.

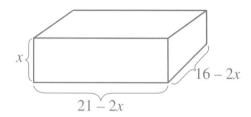

$16 - 2x$

$21 - 2x$

x

❸ Letting $V(x)$ represent the volume of the box, you have:

$$V(x) = (21 - 2x)(16 - 2x)x$$

❹ Since the length, width, and height of the box must be greater than 0, you have:

$$x > 0 \qquad 21 - 2x > 0 \qquad 16 - 2x > 0$$
$$21 > 2x \qquad 16 > 2x$$
$$x < 10.5 \qquad x < 8$$

So the domain of $V(x)$ must be $0 < x < 8$.

❺ Expand $V(x)$ and simplify.

$$V(x) = 4x^3 - 74x^2 + 336x$$

❻ Since you are trying to find a maximum volume, calculate $V'(x)$ and set it equal to 0.

$$V'(x) = 12x^2 - 148x + 336$$
$$0 = 12x^2 - 148x + 336$$

❼ Factor and then solve for x.

$$0 = 4(3x^2 - 37x + 84)$$
$$0 = 4(x - 3)(3x - 28)$$

$x = 3$ or $x = \dfrac{28}{3}$ not in domain of $V(x)$

❽ Since you are locating a maximum in a closed interval [0,8], find the value of V at the endpoints and at the critical numbers in that interval.

$x = 3$

$V(x) = (21 - 2x)(16 - 2x)x$

$V(0) = 0$

$V(3) = (21 - 6)(16 - 6)3 = 450$

$V(8) = (21 - 16)(16 - 16)8 = 0$

❾ $V(3) = 450$ is the maximum box volume.

From each corner, cut squares of 3 inches.

Cylindrical Can Construction Problem

A right circular cylinder has a volume of 2π cubic inches. Find the can dimensions that require the least amount of aluminum to be used in the can's construction.

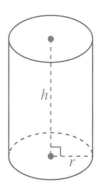

❶ You are looking for the least (that is, minimum) total surface area of the can.

Let A = area function, r = base radius, and h = can height.

$A = 2\pi r^2 + 2\pi rh$

❷ You need to have the area function A in terms of just one variable. Use the given volume to find a relationship between h and r.

$V = \pi r^2 h$

$2\pi = \pi r^2 h$

$\dfrac{2}{r^2} = h$

❸ Substitute $h = \dfrac{2}{r^2}$ into the area function A.

$$A = 2\pi r^2 + 2\pi rh$$

$$A = 2\pi r^2 + 2\pi r \left(\dfrac{2}{r^2} \right)$$

$$A(r) = 2\pi r^2 + \dfrac{4\pi}{r}$$

❹ Find $A'(r)$.

> **TIP**
>
> Instead of using the Quotient Rule to find the last term's derivative, bring the r up to the top and use the Power Rule.

$$A(r) = 2\pi r^2 + \dfrac{4\pi}{r}$$

$$A(r) = 2\pi r^2 + 4\pi r^{-1}$$

$$A'(r) = 4\pi r - 4\pi r^{-2}$$

$$A'(r) = 4\pi r - \dfrac{4\pi}{r^2}$$

❺ Set $A'(r) = 0$ and then solve for r.

> **TIP**
>
> Multiply all terms by r^2 to get rid of the denominator in the last term of the equation.

$$A'(r) = 4\pi r - \dfrac{4\pi}{r^2}$$

$$0 = 4\pi r - \dfrac{4\pi}{r^2}$$

$$0 = 4\pi r^3 - 4\pi$$

$$0 = 4\pi \left(r^3 - 1 \right)$$

$$r = 1$$

The radius of the base is 1.

6 Verify that $r = 1$ yields a minimum value for $A(r)$.
This time, let's use the second derivative test.

$A'(r) = 4\pi r - 4\pi r^{-2}$

$A''(r) = 4\pi + 8\pi r^{-3}$

$A''(r) = 4\pi + \dfrac{8\pi}{r^3}$

$A''(1) = 4\pi + \dfrac{8\pi}{1^3} = 12\pi$

Since $A''(1) > 0$, graph of $A(r)$ is concave upward at $r = 1$, $A(r)$ will have a minimum value at $r = 1$.

7 Find value of h, using $r = 1$.

Note: The result says that the can's height and base radius should be the same number.

$h = \dfrac{2}{r^2}$

$h = \dfrac{2}{1^2}$

$h = 2$

So $r = 1$ and $h = 2$ results in a can with the desired volume, yet the minimum (least) cost to construct.

Bus Company Fare Problem

A bus company currently carries an average of 8,000 riders daily. In anticipation of a fare increase, the bus company conducts a survey of its riders revealing that for each 5¢ increase in the fare, the company will lose an average of 800 riders daily. What fare should the company charge in order to maximize its fare revenue?

1 Let $R(f)$ be the daily revenue function, for which f = the number of 5¢ fare increases. This is shown at right.

$$R(f) = \overbrace{(20 + 5f)}^{\substack{\text{for each 5 cent} \\ \text{fare increase}}}\overbrace{(8000 - 800f)}^{\substack{\text{will lose} \\ \text{800 riders}}}$$

$$R(f) = -4{,}000f^2 + 24{,}000f + 160{,}000$$

2 Find $R'(f)$.

$$R'(f) = -8{,}000f + 24{,}000$$

3 Set $R'(f) = 0$ and solve for f.

$$0 = -8{,}000f + 24{,}000$$
$$8{,}000f = 24{,}000$$
$$f = 3$$

Increasing the fare 3 times gives you $3(5¢) = 15¢$. The new fare, therefore, should be $20 + 5(3) = 35¢$.

4 Verify that $f = 3$ gives the company its maximum revenue.

$$R'(f) = -8{,}000f + 24{,}000$$
$$R''(f) = -8{,}000$$

Since $R''(f) < 0$, the graph of $R(f)$ is concave downward, $f = 3$ results in a **maximum.**

In Chapter 7, you found derivatives such as $\frac{dy}{dy}, \frac{dx}{dt}, \frac{dV}{dt}$ by using implicit differentiation. You will use the same process in this section when two or more related variables are changing with respect to the same third variable—here, time.

Conical Water Tank Problem

A conical tank (with its vertex down) is 8 feet tall and 6 feet across its diameter. If water is flowing into the tank at the rate of 2 ft.³/min., find the rate at which the water level is changing at the instant the water depth is $\frac{2}{3}$ ft.

❶ Let h = the depth of the water in the tank and let r = radius of the circular surface to the water at that time.

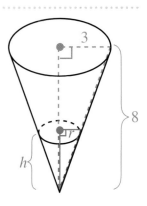

❷ Write a formula for the volume of the water in the tank for a given height and radius.

$$V = \frac{1}{3}\pi r^2 h$$

❸ Since you need to have the volume function in terms of just one variable, find a relationship between h and r by using the similar triangles in the figure at right.

The red right triangle is similar to the blue right triangle.

$$\frac{3}{8} = \frac{r}{h}$$

④ Cross-multiply and then solve for r in terms of h; you are given $\dfrac{dh}{dt}$, so you need to have the volume formula in terms of just h.

$$3h = 8r$$

$$\frac{3h}{8} = r$$

⑤ Substitute $\dfrac{3h}{8} = r$ into the volume formula.

$$V = \frac{1}{3}\pi r^2 h$$

$$V(h) = \frac{1}{3}\pi\left(\frac{3h}{8}\right)^2 h$$

$$V(h) = \frac{1}{3}\pi \cdot \frac{9h^2}{64} \cdot h$$

$$V(h) = \frac{3\pi}{64}h^3$$

⑥ Differentiate, treating V and h as functions of time, t.

$$\frac{dV}{dt} = \frac{3\pi}{64} \cdot 3h^2 \frac{dh}{dt}$$

⑦ The given data is $\dfrac{dV}{dt} = 2$, and $h = \dfrac{2}{3}$. Substitute into the derivative and then solve for $\dfrac{dh}{dt}$.

$$\frac{dV}{dt} = \frac{3\pi}{64} \cdot 3h^2 \frac{dh}{dt}$$

$$2 = \frac{9\pi}{64}\left(\frac{2}{3}\right)^2 \frac{dh}{dt}$$

$$2 = \frac{\pi}{16}\frac{dh}{dt}$$

$$\frac{32}{\pi} = \frac{dh}{dt}$$

The water depth is changing at a rate of $\dfrac{32}{\pi}$ ft./min.

Light Pole and Shadow Problem

A 5-foot-tall woman walks at a rate of 4 feet per second away from a 12-foot-tall pole with a light on top of it.

PART A

Find the rate at which the tip of the woman's shadow is moving away from the base of the light pole.

1 Let w = the distance from the woman to the light pole and let L = the distance from the tip of the shadow to the light pole.

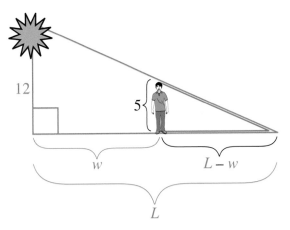

2 Using similar right triangles, write a relationship between L and w.

$$\frac{12}{L} = \frac{5}{L - w}$$

3 Cross-multiply and then solve for L in terms of w.

$$12(L - w) = 5L$$
$$12L - 12w = 5L$$
$$7L = 12w$$
$$L = \frac{12}{7} w$$

❹ Differentiate and again treat all variables as some function of time, t.

$$\frac{dL}{dt} = \frac{12}{7}\frac{dw}{dt}$$

❺ Substitute $\frac{dw}{dt} = 4$ and then solve for $\frac{dL}{dt}$.

$$\frac{dL}{dt} = \frac{12}{7} \cdot 4$$

Note: *It may seem strange, but the rate at which her shadow is moving away from the pole is independent of her distance from the pole!*

$$\frac{dL}{dt} = \frac{48}{7}$$

The tip of her shadow is moving away from the pole at a rate of $\frac{48}{7}$ ft./ sec.

PART B

Find the rate at which the length of the woman's shadow is changing.

❶ Let w = the distance from the woman to the light pole and let L = the length of the shadow (the distance from the woman to the tip of the shadow).

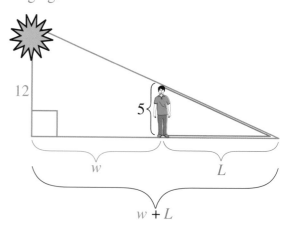

❷ Relate w and L using the red and blue similar right triangles.

$$\frac{12}{w + L} = \frac{5}{L}$$

❸ Solve for L in terms of w.

$$12L = 5(w + L)$$
$$12L = 5w + 5L$$
$$7L = 5w$$
$$L = \frac{5}{7}w$$

❹ Differentiate with respect to time, t.

$$\frac{dL}{dt} = \frac{5}{7}\frac{dw}{dt}$$

❺ Substitute $\frac{dw}{dt} = 4$ and then solve for $\frac{dL}{dt}$.

$$\frac{dL}{dt} = \frac{5}{7} \cdot 4$$
$$\frac{dL}{dt} = \frac{20}{7}$$

The length of her shadow is changing at a rate of $\frac{20}{7}$ ft./sec.

Position, Velocity, and Acceleration

In previous chapters, you found that if $s(t)$ = the position function, then $s'(t) = v(t)$ = the velocity function and $s''(t) = v'(t) = a(t)$ = the acceleration function.

Rocket Problem

A small toy rocket is shot into the air from the top of a tower. Its position, s, in feet, t seconds after liftoff, is given by the equation $s(t) = -16t^2 + 32t + 240$.

PART A

Find the velocity of the rocket 2 seconds after liftoff.

① Starting with the position function $s(t)$, find the velocity function $v(t)$.

$$s(t) = -16t^2 + 32t + 240$$
$$s'(t) = v(t) = -32t + 32$$

② Find $v(2)$.

$$v(t) = -32t + 32$$
$$v(2) = -32(2) + 32$$
$$v(2) = -32$$

After 2 seconds, the rocket's velocity is -32 feet/second. (The negative velocity indicates that the rocket is moving downward.)

PART B

For how many seconds will the rocket be in the air?

Find the time at which the rocket lands on the ground—that is, when $s(t) = 0$.

$$s(t) = -16t^2 + 32t + 240$$
$$0 = -16t^2 + 32t + 240$$
$$0 = -16(t^2 + 2t - 15)$$
$$0 = -16(t - 5)(t + 3)$$
$$t = 5 \qquad \cancel{t = -3}$$

The rocket is in the air for 5 seconds.

PART C

With what velocity will the rocket hit the ground?

1 Start with the velocity function $v(t)$.

$$v(t) = -32t + 32$$

2 Find $v(5)$.

Note: We used t = 5, since that's the time when the rocket hits the ground.

$$v(5) = -32 \cdot 5 + 32$$
$$v(5) = -160 + 32$$
$$v(5) = -128$$

The rocket hits the ground with a velocity of -128 feet/second. Again, the negative indicates that the rocket is moving downward.

PART D

How many seconds after liftoff will the rocket reach its maximum height?

You need to find when the rocket stops moving—that is, when $v(t) = 0$.

$$v(t) = -32t + 32$$
$$0 = -32t + 32$$
$$-32t = 32$$
$$t = 1$$

The rocket reaches its maximum height after 1 second.

PART E

Find the maximum height reached by the rocket.

1 Start with position function, $s(t)$.

$$s(t) = -16t^2 + 32t + 240$$

❷ Find $s(1)$.

$$s(1) = -16(1)^2 + 3291) + 240$$
$$s(1) = -16 + 32 + 240$$
$$s(1) = 256$$

The maximum height reached by the rocket is 256 feet.

Particle Moving Along a Straight Line Problem

This type of problem is also known as a "rectilinear motion" problem. A particle moves along the *x*-axis so that its *x*-coordinate at time t (seconds) is given by the position function: $x(t) = 3t^4 - 28t^3 + 60t^2$.

PART A

At what time is the particle at rest?

❶ The particle is at rest when its velocity is 0. First, find the velocity function.

$$x(t) = 3t^4 - 28t + 60t^2$$
$$x'(t) = v(t) = 12t^3 - 84t^2 + 120t$$

❷ Set $v(t) = 0$ and then solve for t.

$$0 = 12t^3 - 84t^2 + 120t$$
$$0 = 12t(t^2 - 7t + 10)$$
$$0 = 12t(t - 2)(t - 5)$$
$$t = 0, t = 2, \text{ and } t = 5$$

The particle is at rest at 0, 2, and 5 seconds.

PART B

During what time intervals is the particle moving to the left? During what time intervals is the particle moving to the right?

Create a first derivative chart, using the numbers obtained in Part A.

$x'(t) = v(t) = 12t^3 - 84t^2 + 120t$		
$0 < t < 2$	$2 < t < 5$	$t > 5$
$v(1) > 0$	$v(3) < 0$	$v(6) > 0$
$\rightarrow$	$\leftarrow$	$\rightarrow$
right	left	right

The particle is moving right when $0 < t < 2$ and when $t > 5$.

The particle is moving left when $2 < t < 5$.

> **TIP**
>
> When the velocity is negative, the particle is moving left.
>
> When the velocity is positive, the particle is moving right.

PART C

Find the total distance traveled by the particle in the first 5 seconds.

1 Start with the position function, $x(t)$.

$$x(t) = 3t^4 - 28t^3 + 60t^2$$

2 Using the zeros of $v(t)$, $t = 0$, $t = 2$, and $t = 5$, find the value of $x(t)$ for each of these times.

$x(0) = 0$
$x(2) = 64$
$x(5) = -125$

3 Find the distance traveled in each time interval, and then find the sum of these two distances.

From 0 to 2 seconds: $64 - 0 = 64$ units
From 2 to 5 seconds: $64 - (-125) = 189$ units
Total distance traveled in first 5 seconds is $64 + 189 = 253$ units.

PART D

What is the particle's acceleration at $t = 1$?

❶ Start with the velocity function $v(t)$.

$$v(t) = 12t^3 - 84t^2 + 120t$$

❷ Find $a(t)$, the acceleration function.

$$v'(t) = a(t) = 36t^2 - 168t + 120$$

❸ Find $a(1)$.

$$a(1) = 36(1)^2 - 168(1) + 120$$
$$a(1) = -12$$

The particle's acceleration at $t = 1$ is -12 *ft*/sec², with the negative indicating that the particle is slowing down or "decelerating."

PART E

At what time is the particle moving with constant velocity?

❶ Begin with the acceleration function, $a(t)$.

$$a(t) = 36t^2 - 168t + 120$$

❷ The particle moves at constant velocity when it is not accelerating, so set $a(t) = 0$.

$$0 = 36t^2 - 168t + 120$$
$$0 = 12(3t^2 - 14t + 10)$$

....

❸ Since the equation factors no further, use the Quadratic Formula to find the values of t.

$$t = \frac{-(-14) \pm \sqrt{(-14)^2 - 4(3)(10)}}{2(3)}$$

$$t = \frac{14 \pm \sqrt{196 - 120}}{6}$$

$$t = \frac{14 \pm 2\sqrt{19}}{6}$$

$$t = \frac{7 \pm \sqrt{19}}{3}$$

The particle is moving with constant velocity at approximately 3.79 seconds and 0.88 seconds.

TIP

For the quadratic equation $ax^2 + bx + c = 0$, the solution is

$$x = \frac{-b \pm \sqrt{b^2 - 4ac}}{2a}$$

Introduction to the Integral

This chapter is the first of three chapters that deal with the process of starting with the derivative of a function and working backward to get the original function, called the **antiderivative**. This process is known as **integration**. This chapter covers both the indefinite and definite integrals, along with their properties, as well as the First and Second Fundamental Theorems of Calculus and the Mean Value Theorem.

Antiderivatives: Differentiation
  versus Integration 195

The Indefinite Integral and
  Its Properties 197

Common Integral Forms 201

First Fundamental Theorem
  of Calculus 203

The Definite Integral and Area 205

Second Fundamental Theorem of
  Calculus 209

Antiderivatives: Differentiation versus Integration

The process of finding a function from which a given derivative is derived is known as **antidifferentiation**, or **integration**. This section introduces that relationship and covers the indefinite integral and its properties.

Definition of an Antiderivative

A function, F, is called an **antiderivative** of function f on an interval if $F'(x) = f(x)$ for all x in that interval.

Let $F(x) = x^3 - 7x + 6$; then $F'(x) = f(x) = 3x^2 - 7$.

❶ One antiderivative of $f(x) = 3x^2 - 7$ is the function $F(x)$ at right.

$$F(x) = x^3 - 7x + 6$$

❷ A second antiderivative of $f(x) = 3x^2 - 7$ is the function $F(x)$ at right.

$$F(x) = x^3 - 7x - 15$$

❸ In each case above, $F'(x) = f(x)$. So it appears that a given function $f(x)$ has an infinite number of antiderivatives, $F(x)$, all differing from each other by just a constant.

You, therefore, write the most general antiderivative of $f(x) = 3x^2 - 7$ as $F(x) = x^3 - 7x + c$, where c is just some constant.

Finding Some Antiderivatives

1 Find the antiderivative of $f(x) = \cos x$.

$F(x) = \sin x + c$ **because** $F'(x) = f(x)$

2 Find the antiderivative of $f(x) = \frac{1}{x}$.

$F(x) = \ln x + c$ **because** $F'(x) = f(x)$.

3 Find the antiderivative of $f(x) = e^x$.

$F(x) = e^x + c$ **because** $F'(x) = f(x)$.

The Indefinite Integral and Its Properties

This section introduces the indefinite integral (an antiderivative) of a function along with its properties. The section also includes some examples of finding indefinite integrals.

Definition of the Indefinite Integral

The **indefinite integral** of a function $f(x)$, written as $\int f(x)\,dx$, is the set of all antiderivatives of the function $f(x)$.

In the expression $\int f(x)\,dx$:

$\int$ is the integral symbol.

"$f(x)$" is called the **integrand**.

"dx" tells you that the variable of integration is x.

$\int f(x)\,dx$ is read "the integral of f of x with respect to x."

FINDING SOME INDEFINITE INTEGRALS

1. Because $\dfrac{d}{dx}(x^3 - 7x) = 3x^2 - 7 \Rightarrow$

$$\int (3x^2 - 7)\,dx = x^3 - 7x + c$$

2. Since $\dfrac{d}{dx}(\sin x) = \cos x \Rightarrow$

$$\int \cos x\,dx = \sin x + c$$

3. Because $\dfrac{d}{dx}(e^{2x}) = 2e^{2x} \Rightarrow$

$$\int (2e^{2x})\,dx = e^{2x} + c$$

4 Since $\frac{d}{dx}(\ln x) = \frac{1}{x} \Rightarrow$

$$\int \left(\frac{1}{x}\right) dx = \ln|x| + c$$

> **TIP**
>
> The $|x|$ is there since you can't take the natural log of a negative number.

FINDING A PARTICULAR ANTIDERIVATIVE

Find the particular antiderivative of $f'(x) = 3x^2 - 7$ that satisfies the condition $f(1) = 3$.

You need to find a specific or "particular" value of c for the antiderivative of $3x^2 - 7$.

1 $f(x) = \int f'(x)\, dx$

$f(x) = \int \left(3x^2 - 7\right) dx$

$f(x) = x^3 - 7x + c$

This was shown in the preceding section.

2 Since $f(1) = 3$, when $x = 1$, $f(x) = 3$.

$3 = (1)^3 - 7(1) + c$

$9 = c$

3 Since you found a particular (or specific) value of c, you have found a "particular" antiderivative of $f'(x) = 3x^2 - 7$.

$f(x) = x^3 - 7x + 9$ is the particular antiderivative of $f'(x) = 3x^2 - 7$.

> **TIP**
>
> **Continuity Implies Integrability**
> If a function f is continuous on the closed interval $[a,b]$, then f is also integrable on $[a,b]$. (The term **integrable** means that you are able to integrate it.)

FINDING A FUNCTION FROM ITS SECOND DERIVATIVE

Find the function $f(x)$ for which $f''(x) = 32$, $f'(1) = 36$, and $f(1) = 16$.

---

① Find $f'(x)$ from the given data $f''(x) = 32$.

$$f'(x) = \int f''(x)\,dx$$
$$f'(x) = \int 32\,dx$$
$$f'(x) = 32x + c_1$$

---

② You are given $f'(1) = 36$; use this to find the value of c_1.

$$f'(1) = 32(1) + c_1$$
$$36 = 32 + c_1$$
$$4 = c_1$$

Therefore, $f'(x) = 32x + 4$.

---

③ Find $f(x)$.

Note: *A second constant, c_2, is used here. The constant c_1 is from the first integration step. We cannot assume that these are equal, so they need to be labeled separately.*

$$f(x) = \int f'(x)\,dx$$
$$f(x) = \int (32x + 4)\,dx$$
$$f(x) = 16x^2 + 4x + c_2$$

---

④ You are told that $f(1) = 16$; use this fact to find the value of c_2.

$$f'(1) = 16(1)^2 + 4(1) + c_2$$
$$16 = 16 + 4 + c_2$$
$$-4 = c_2$$

Thus, $f(x) = 16x^2 + 4x - 4$.

Properties of Indefinite Integrals

If f and g are continuous functions and defined on the same interval and K is some constant, then the following properties apply:

❶ Integral of dx: $\int dx = x + c$

$$\int 5\,dx = 5x + c$$

❷ Integral of a constant: $\int k\,dx = kx + c$

$$\int \frac{1}{2}\,dx = \frac{1}{2}x + c$$

❸ Integral of constant times a function:

$\int k \cdot f(x)\,dx = 5 \cdot \int f(x)\,dx$, where $4c = m$ is just another constant.

$$\int 4(3x^2 - 7)\,dx = 4\int (3x^2 - 7)\,dx$$
$$= 4(x^3 - 7x + c)$$
$$= 4x^3 - 28x + 4c$$
$$= 4x^3 - 28x + m$$

Another way to deal with the constant is as follows:

$$\int 4(3x^2 - 7)\,dx = 4\int (3x^2 - 7)\,dx$$
$$= 4(x^3 - 7x) + c$$
$$= 4x^3 - 28x + c$$

Just integrate everything and put a $+ c$ at the end.

❹ Integral of the sum/difference of functions:

$$\int (f(x) \pm g(x))\,dx = \int f(x)\,dx \pm \int g(x)\,dx$$

$$\int \left(\cos x + \frac{1}{x} \right) dx = \int \cos x\,dx + \int \frac{1}{x}\,dx$$
$$= \sin x + \ln|x| + c$$

Note that only one "+ c" was written; if you used separate "+ constant" for each function, their sum would just be another constant anyway.

To create an integral formula from a known derivative formula, just write the formula in "reverse," adding the correct integral notation and the "+ c." For example, since $\frac{d}{dx}(\sin x) = \cos x$, you can also write that $\int \cos x \, dx = \sin x + c$. The following integral formulas were created by just reading an existing differentiation formula in reverse.

POWER

Use the formula below to integrate some power of a variable. If you were to differentiate the right side, you would end up with the left side.

$$\int x^3 \, dx = \frac{x^4}{4} + c \text{ or } \frac{1}{4}x^4 + c$$

$$\int \frac{1}{2\sqrt{x}\,dx} = \int \frac{1}{2} x^{-1/2} \, dx = \frac{1}{2}\left[\frac{x^{1/2}}{\frac{1}{2}}\right] + c = x^{1/2} + c = \sqrt{x} + c$$

$$\int x^n \, dx = \frac{x^{n+1}}{n+1} + c, \text{ for } n \neq -1$$

POLYNOMIAL

Using a combination of the properties listed in previous sections and the Power Rule listed above, you can find the integral of a polynomial as follows:

$$\int \left(a_n x^n + a_{n-1} + \ldots + a_2 x^2 + a_1 x^1 + a_0\right) dx$$

$$= a_n \frac{x^{n+1}}{n+1} + a_{n-1}\frac{x^n}{n} + \ldots a_2 \frac{x^3}{3} + a_1 \frac{x^2}{2} + a_0 x + c$$

$$\int \left(3x^2 - 6x + 5\right) dx$$

$$= 3 \cdot \frac{x^3}{3} - 6 \cdot \frac{x^2}{2} + 5x + c$$

$$= x^3 - 3x^2 + 5x + c$$

NATURAL LOGARITHM

Integrating the expression $\frac{1}{x}$ is just a matter of using the derivative on lnx in reverse. Since $\frac{d}{dx}\ln x = \frac{1}{x}$ the following formula must be true.

$$\int \frac{1}{x}\,dx = \ln|x| + c$$

$$\int \frac{3}{x}\,dx = 3\int \frac{1}{x}\,dx = 3\ln|x| + c$$

Note: *Since you cannot find the natural log of a negative number, the absolute value* $|n|x|$ *is used.*

EXPONENTIAL

The following derivative rules come from their appropriate derivative counterparts for exponential functions found in Chapter 6.

$$\int e^x\,dx = e^x + c \text{ and } \int a^x\,dx = \frac{a^x}{\ln a} + c$$

$$\int 5e^x\,dx = 5\int e^x\,dx = 5e^x + c$$

$$\int 2^x\,dx = \frac{2^x}{\ln 2} + c$$

TRIGONOMETRIC: COSINE AND SINE

The following integral formulas follow directly from their derivative counterparts found in Chapter 5.

$$\int \cos x\,dx = \sin x + c$$
$$\int \sin x\,dx = -\cos x + c$$

$$\int (2\cos x + 3\sin x)\,dx$$
$$= \int 2\cos x\,dx + \int 3\sin x\,dx$$
$$= 2\int \cos x\,dx + 3\int \sin x\,dx$$
$$= 2\sin x + 3(-\cos x) + c$$
$$= 2\sin x - 3\cos x + c$$

TRIGONOMETRIC: SOME OF THE OTHERS

Looking back at the trigonometric derivative formulas in Chapter 6, you can see that the following integral formulas result from the process of antidifferentiation.

$$\int \sec^2 x\,dx = \tan x + c$$
$$\int \csc^2 x\,dx = -\cot x + c$$
$$\int \sec x\tan x\,dx = \sec x + c$$
$$\int \csc x\cot x\,dx = -\csc x + c$$

$$\int (\sec^2 x + \sec x\tan x)\,dx$$
$$= \int \sec^2 x\,dx + \int \sec x\tan x\,dx$$
$$= \tan x + \sec x + c$$
$$= \tan x + \sec x + c$$

If the function f is continuous on the closed interval $[a,b]$ and F is an antiderivative of f (that is, $F'(x) = f(x)$) on the interval $[a,b]$, then $\int_a^b f(x)\,dx = F(b) - F(a)$.

Another way to write the final result is $= \left[F(x)\right]_a^b = F(b) - F(a)$. In other words, it says "after finding the antiderivate, $F(x)$, find the value at the top limit of integration, $F(b)$, then find the value at the bottom limit of integration, $F(a)$, and then find their difference, $F(b) - F(a)$."

The a and b on the integral sign $\int_a^b f(x)\,dx$ are called the **limits of integration**, and the dx indicates that a and b are x values; thus the function $f(x)$ being integrated must be a function of x.

The expression $\int_a^b f(x)\,dx$ is called a **definite integral**.

Note: The definite integral $\int_a^b f(x)\,dx$ represents a number (a definite value), while the indefinite integral f(x)dx represents a family of functions (remember the "+ c"), and not a definite, or specific, function.

....

EVALUATE A DEFINITE INTEGRAL: EXPONENTIAL FUNCTION

Evaluate $\int_0^1 e^x\,dx$.

$\int_a^b f(x)\,dx = \left[F(x)\right]_a^b = F(b) - F(a)$

follow the pattern above for problem below

$\int_0^1 e^x\,dx = \left[e^x\right]_0^1 = e^1 - e^0 = e - 1$

....

EVALUATE A DEFINITE INTEGRAL: TRIGONOMETRIC FUNCTION

Evaluate $\int_0^{\pi/6} \cos x\,dx$.

$\int_0^{\pi/6} \cos x\,dx = \left[\sin x\right]_0^{\pi/6} = \sin\left(\frac{\pi}{6}\right) - \sin(0) = \frac{1}{2} - 0 = \frac{1}{2}$

First Fundamental Theorem of Calculus *(continued)*

EVALUATE A DEFINITE INTEGRAL: POLYNOMIAL FUNCTION

Evaluate $\int_{-1}^{2} (2x - 3x^2)\, dx$.

❶ Find the integral with limits written on the bracket.

$$\int_{-1}^{2} (2x - 3x^2)\, dx$$
$$= \left[x^2 - x^3 \right]_{-1}^{2}$$

· ·

❷ With $F(x) = x^2 - x^3$, find $F(2) - F(-1)$.

$$= \underbrace{\left(2^2 - 2^3 \right)}_{F(2)} - \underbrace{\left((-1)^2 - (-1)^3 \right)}_{F(-1)}$$
$$= (4 - 8) - (1 + 1)$$
$$= -6$$

One of the applications of the definite integral is finding the area of a region bounded by the graphs of two functions.

Let f be a continuous function on $[a,b]$ for which $f(x) \geq 0$ for all x in $[a,b]$. Let R be the region bounded by the graphs of $y = f(x)$ and the x-axis and the vertical lines $x = a$ and $x = b$. Then the area, A_R, of the region is given by $A_R = \int_a^b f(x)\,dx$.

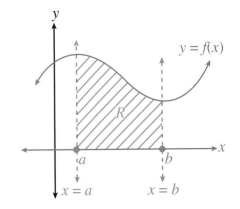

Area of a Bounded Region: Linear Function

Find the area of the region bounded by the graph of $y = 2x$, $y = 0$ (the x-axis), and the lines $x = 0$ (the y-axis) and $x = 3$.

❶ Sketch a diagram of the bounded region.

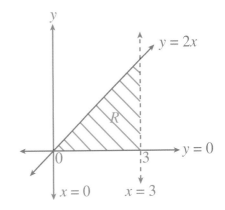

❷ Set up the integral with $a = 0$, $b = 3$, and $f(x) = 2x$.

$$A_R = \int_a^b f(x)\,dx$$
$$= \int_0^3 2x\,dx$$

❸ Evaluate the integral.

$$= \left[\frac{2x^2}{2}\right]_0^3 = \left[x^2\right]_0^3 = 3^2 - 0^2 = 9$$

Therefore, $A_R = 9$.

❹ You could have just found the area of the triangle with a base of 3 and a height of 6.

$$A = \frac{1}{2}bh = \frac{1}{2} \cdot 3 \cdot 6 = 9$$

AREA OF A BOUNDED REGION: TRIGONOMETRIC FUNCTION

Find the area of the region bounded by the graphs of $y = \cos x$, $y = 0$, and $x = 0$.

❶ Sketch a diagram; note that $\cos x = 0$ when $x = \frac{\pi}{2}$.

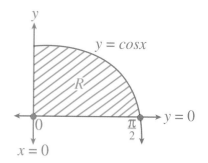

❷ Set up the appropriate integral representing the area of the given region.

$$A_R = \int_0^{\pi/2} \cos x\, dx$$

❸ Evaluate the integral.

$$= \left[\sin x\right]_0^{\pi/2} = \sin\left(\frac{\pi}{2}\right) - \sin(0) = 1 - 0 = 1$$

Some Properties of the Definite Integral

❶ If the function f is defined at $x = a$, then $\int_a^a f(x)\,dx = 0$.

It simply states that the area from $x = a$ to $x = a$ is 0.

There is no work required; the area is just 0.

$$\int_5^5 \left(\ln x + e^x + \sin x\right)dx = 0$$

❷ If the function f is integrable on $[a,b]$, then

$\int_a^b f(x)\,dx = -\int_b^a f(x)\,dx$. This makes sense, since by

switching the order of the limits of integration, you just

switch the order of the subtraction when you evaluate

the integral.

$$\int_5^2 f(x)\,dx = -\int_2^5 f(x)\,dx$$

❸ For $a < b < c$, if the function f is integrable on $[a,b]$, $[b,c]$,
and $[a,c]$, then $\int_a^c f(x)\,dx = \int_a^b f(x)\,dx + \int_b^c f(x)\,dx$.

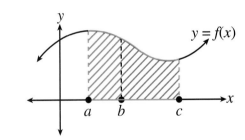

In terms of area,

$$\underbrace{\int_a^c f(x)\,dx}_{\substack{area\ of\ green \\ region}} = \underbrace{\int_a^b f(x)\,dx}_{\substack{area\ of\ red \\ region}} + \underbrace{\int_b^c f(x)\,dx}_{\substack{area\ of\ blue \\ region}}$$

4 If the function f is integrable on $[a,b]$

$$\int_1^2 12x^3\,dx = 12 \cdot \int_1^2 x^3\,dx = 12\left[\frac{x^4}{4}\right]_1^2 = 12\left(\frac{16}{4} - \frac{1}{4}\right) = 12\left(\frac{15}{4}\right) = 45$$

and k is a constant,

then $\int_a^b k \cdot f(x)\,dx = k \cdot \int_a^b f(x)\,dx$.

5 If the functions f and g are integrable on $[a,b]$, then

$$\int_a^b \big(f(x) \pm g(x)\big)\,dx = \int_a^b f(x)\,dx \pm \int_a^b g(x)\,dx.$$

$$\int_1^e \left(\frac{1}{x} + e^x\right)dx$$
$$= \int_1^e \frac{1}{x}\,dx + \int_1^e e^x\,dx$$
$$= \ln(e) - \ln(1) + e^e - e^1$$
$$= 1 - 0 + e^e - e^1$$
$$= 1 + e^e - e^1$$

Let f be a function that is continuous on $[a,b]$, and let x be any number in $[a,b]$; then

$$\frac{d}{dx}\int_a^x f(t)\,dt = f(x).$$

Remember, if you let $F(x) = \int_a^x f(t)\,dt$, then the theorem is just saying that $F'(x) = f(x)$. For example,

$$\frac{d}{dx}\int_2^x \underbrace{\left(\ln t + e^t\right)}_{f(t)}\,dt = \underbrace{\ln x + e^x}_{f(x)}.$$ Just replace $f(t)$ with $f(x)$.

EXAMPLE 1

Find $\frac{d}{x}\int_3^x \left(t^2 - 2t\right)\,dt$.

$$\frac{d}{dx}\int_3^x \underbrace{\left(t^2\right) - 2t}_{f(t)}\,dt = \underbrace{x^2 - 2x}_{f(x)} \text{ or just } x^2 - 2x$$

EXAMPLE 2

Find $\dfrac{d}{dx}\displaystyle\int_5^x \left(\sin t + e^t - t^3\right)\,dt = \sin x + e^x - x^3$

or just $\sin x + e^x - x^3$.

EXAMPLE 3

Find $\dfrac{d}{dx}\displaystyle\int_{2\sqrt{3}}^x \dfrac{\sqrt{t} + \ln t}{t^3}\,dt$.

$$\frac{d}{dx}\int_{2\sqrt{3}}^x \frac{\sqrt{t} + \ln t}{t^3}\,dt = \frac{\sqrt{x} + \ln x}{x^3} \text{ or } \frac{\sqrt{x} + \ln x}{x^3}$$

SECOND FUNDAMENTAL THEOREM OF CALCULUS: TOP LIMIT IS A FUNCTION OF *X*

What happens when the top limit of integration is some function of *x*, other than *x* itself?

If u is a function of x, then $\dfrac{d}{dx}\displaystyle\int_{a}^{u} f(t)\,dt = f(u) \cdot du$.

❶ Find $\dfrac{d}{dx}\displaystyle\int_{4}^{x^3} \left(\ln t + e^t \right) dt$.

$$\frac{d}{dx}\int_{4}^{x^3} \underbrace{\left(\ln t + e^t \right)}_{f(t)}\,dt = \Big[\underbrace{\ln\left(x^3\right) + e^{x^3}}_{f\left(x^3\right)} \Big] \cdot \underbrace{3x^2}_{\substack{\text{der. of} \\ x^3}}$$

or just $\Big[\ln x^3 + e^{x^3} \Big] \cdot 3x^2$

❷ Find $\dfrac{d}{dx}\displaystyle\int_{\pi}^{\ln x} \left(t^2 + \sin t \right) dt$.

$$\frac{d}{dx}\int_{\pi}^{\ln x} \underbrace{\left(t^2 + \sin t \right)}_{f(t)}\,dt = \Big[\underbrace{\left(\ln x\right)^2 + \sin\left(\ln x\right)}_{f\left(\ln x\right)} \Big] \cdot \underbrace{\frac{1}{x}}_{\substack{\text{der. of} \\ \ln x}}$$

11

Techniques of
Integration

This chapter introduces many techniques of
integration, the process of finding the integral
of a function. A lot of these techniques depend upon
your being able to recognize a pattern in the way
the function is, or can be, written—such as in an
exponential function or some power of a function,
or a function whose integral will result in a
logarithmic function.

   This chapter introduces integration techniques
such as integrals involving trigonometric functions,
or integrals that result in inverse trigonometric
functions. Sometimes multiple techniques are
required to integrate a given function. The use of
algebraic substitution covers some integrals that do
not seem to fit any other pattern of integration. The
chapter concludes with solving some differential
equations.

Power Rule: Simple and General 212

Integrals of Exponential Functions. . . 220

Integrals That Result in a Natural
   Logarithmic Function 223

Integrals of Trigonometric
   Functions. 226

Integrals That Result in an Inverse
   Trigonometric Function. 232

Combinations of Functions and
   Techniques 235

Algebraic Substitution. 237

Solving Variables Separable
   Differential Equations 240

Power Rule: Simple and General

There are two versions of the Power Rule to consider when finding the integral of a function: the **Simple Power Rule**, in which you integrate powers of the term x, and the **General Power Rule**, in which you integrate powers of a function of x.

Simple Power Rule

$$\int x^n \, dx = \frac{x^{n+1}}{n+1} + c, \text{ where } n \neq -1$$

EXAMPLE 1

$\int x^2 \, dx$

❶ Start with the given expression.

$$\int \overbrace{x^2}^{x^n} \, dx$$

❷ Integrate, applying the Simple Power Rule, by increasing the power of x to 3 and then dividing the new term by 3.

$$= \frac{\overbrace{x^3}^{x^{n+1}}}{\underbrace{3}_{n+1}} + c$$

$$= \frac{x^3}{3} + c$$

EXAMPLE 2

$\int \frac{1}{x^2} \, dx$

❶ Rewrite the integrand as a negative power of x.

$$= \int x^{-2} \, dx$$

❷ Apply the Simple Power Rule by adding 1 to the exponent and dividing the new term by the new exponent of -1.

$$= \frac{x^{\overbrace{-1}^{n+1}}}{\underbrace{-1}_{n+1}}$$

❸ Simplify the resulting expression.

$$= -\frac{1}{x} + c$$

EXAMPLE 3

$\int \sqrt[3]{x^2}\, dx$

❶ Rewrite the radical term as an exponential term.

$$\int \sqrt[3]{x^2}\, dx$$
$$= \int x^{2/3}\, dx$$

❷ Apply the Simple Power Rule and simplify the result.

$$= \frac{x^{5/3}}{\frac{5}{3}} + c$$
$$= \frac{3}{5} x^{5/3} + c$$

❸ Rewrite the result as a radical term since the original integrand was a radical term.

$$= \frac{3}{5} \sqrt[3]{x^5} + c$$
$$= \frac{3}{5} x \sqrt[3]{x^2} + c$$

EXAMPLE 4

$$\int \left(\frac{3x^4 - 4x^2 + 7}{x^2} \right) dx$$

❶ Rewrite the original rational expression as three separate rational expressions.

$$\int \left(\frac{3x^4 - 4x^2 + 7}{x^2} \right) dx$$

$$= \int \left(\frac{3x^4}{x^2} - \frac{4x^2}{x^2} + \frac{7}{x^2} \right) dx$$

❷ Simplify each rational expression.

$$= \int \left(3x^2 - 4 + 7x^{-2} \right) dx$$

❸ Break up the expression into three separate integrals.

Remember, you can just move the constants, 3 and 7, outside of the integral sign.

$$= 3\int x^2 \, dx - \int 4 \, dx + 7\int x^{-2} \, dx$$

❹ Evaluate each integral by adding just one "+ c" at the end.

$$= 3\left(\frac{x^3}{3} \right) - 4x + 7\left(\frac{x^{-1}}{-1} \right) + c$$

$$= x^3 - 4x - \frac{7}{x} + c$$

General Power Rule

If u is a function of another variable, say x, then $\int u^n \, du = \frac{u^{n+1}}{n+1} + c$, where $n \neq -1$.

Remember that du is just the derivative of the function u. Another way to view this is

$$\int \left(\text{orig. funct.} \right)^n \cdot \left(\text{der. of orig. funct.} \right) = \frac{\left(\text{orig. funct.} \right)^{n+1}}{n+1} + c$$

EXAMPLE 1

$\int (x^2 + 5)^6 2x dx$

① Identify the u, n, and du for this problem.

$$\int \overbrace{(x^2 + 5)}^{u} {}^{\overset{n}{6}} \overbrace{2x dx}^{du}$$

② Apply the General Power Rule: $\dfrac{\left(\text{orig. funct.}\right)^{n+1}}{n+1} + c$

$$= \dfrac{\overbrace{(x^2 + 5)}^{u} {}^{\overset{n+1}{7}}}{\underbrace{7}_{n+1}} + c$$

$$= \tfrac{1}{7}(x^2 + 5) + c$$

EXAMPLE 2

$\int x^2 (x^3 - 7)^4 dx$

① Although this integral looks like it might be a Simple Power Rule situation, let's move the terms around to be sure.

$$\int x^2 (x^3 - 7)^4 dx$$
$$= \int (x^3 - 7)^4 \cdot x^2 dx$$

② Unfortunately, the $x^2 dx$ term is not quite the derivative of the inside function $x^3 - 7$. Since the derivative of $x^3 - 7 = 3x^2 dx$, you can multiply by 3 inside the integral, and compensate for it by multiplying the outside by $\tfrac{1}{3}$.

$$= \tfrac{1}{3}\underbrace{(x^3 - 7)^4}_{u} \cdot \underbrace{3x^2 dx}_{du}$$

Power Rule: Simple and General *(continued)*

③ Now that the integral fits the General Power Rule pattern, you can just increase the exponent 4 by 1 to a 5, and then divide this term by 5.

$$= \frac{1}{3}\left[\frac{\left(x^3- 7\right)^5}{5}\right]+ c$$

$$= \frac{\left(x^3- 7\right)^5}{15} + c$$

EXAMPLE 3 (USING CHANGE OF VARIABLE OR U-SUBSTITUTION TECHNIQUE)

$$\int x^2\left(x^3- 7\right)^4 dx$$

① After writing the original integrand in a more useful form (it looks like a General Power Rule pattern), you have the expression at right.

$$\int x^2\left(x^3- 7\right)^4 dx$$

$$= \int \left(x^3- 7\right)^4 \cdot x^2 \, dx$$

② Let's try a different approach by using what is called the "change or variable" or "*u*-substitution" technique, letting *u* = the inside function and then proceed as shown at right.

let $u = x^3- 7$, then $du = 3x^2 \, dx$

or $\frac{1}{3} \, du = x^2 \, dx$

Notice that you now have all the terms of the original integrand written in terms of the new variable *u*.

③ Now substitute the *u* terms found in Step 3 for the corresponding parts in Step 2.

$$= \int u^4 \cdot \frac{1}{3} \, du$$

④ Move the $\frac{1}{3}$ out in front of the integral and apply the General Power Rule.

$$= \frac{1}{3} \int u^4 \, du$$

$$= \frac{1}{3} \cdot \frac{u^5}{5} + c$$

$$= \frac{u^5}{15} + c$$

⑤ You need to return to a function in terms of x, not u. So substitute for $u = x^3 - 7$ in the result from Step 4.

$$= \frac{\left(x^3 - 7\right)^5}{15} + c \text{ or just } \frac{\left(x^3 - 7\right)^5}{15} + c$$

EXAMPLE 4 (WITH LIMITS OF INTEGRATION)

$$\int_0^2 \frac{5x^2}{\sqrt{x^3 + 1}} \, dx$$

❶ Rewrite the integrand by bringing the radical term from the denominator up to the numerator as an exponential term instead.

$$\int_0^2 \frac{5x^2}{\sqrt{x^3 + 1}} \, dx$$

$$= \int_0^2 \frac{5x^2}{\left(x^3 + 1\right)^{1/2}} \, dx$$

$$= \int_0^2 5x^2 \left(x^3 + 1\right)^{-1/2} \, dx$$

❷ Move some of the terms so that it looks more like a General Power Rule situation.

$$= \int_0^2 \left(x^3 + 1\right)^{-1/2} \cdot 5x^2 \, dx$$

③ Rewrite the term $5x^2$ as the derivative of the inside function $x3 + 1$ (that is, you need a $3x^2$, not a $5x^2$).

$$= 5 \cdot \int_0^2 (x^3 + 1)^{-1/2} \cdot x^2 \, dx$$

$$= 5 \cdot \frac{1}{3} \cdot \int_0^2 (x^3 + 1)^{-1/2} \cdot 3x^2 \, dx$$

$$= \frac{5}{3} \cdot \int_0^2 \underbrace{(x^3 + 1)^{-1/2}}_{u} \cdot \underbrace{3x^2 \, dx}_{du}$$

④ You are finally ready to apply the General Power Rule.

$$= \frac{5}{3} \left[\frac{(x^3 + 1)^{1/2}}{\frac{1}{2}} \right]_0^2$$

$$= \frac{10}{3} \left[\sqrt{x^3 + 1} \right]_0^2$$

⑤ Plug in the limits of integration to simplify the result.

$$= \frac{10}{3} \left[\sqrt{2^3 + 1} - \sqrt{0^3 + 1} \right]$$

$$= \frac{10}{3} \left(\sqrt{9} - \sqrt{1} \right)$$

$$= \frac{10}{3} (2) = \frac{20}{3}$$

EXAMPLE 5

$$\int_0^{\pi/2} \cos x \sqrt{\sin x} \, dx$$

① First, note that the derivative of $\sin x$ is $\cos x$. Use this fact to set up the integrand in the General Power Rule format.

$$\int_0^{\pi/2} \cos x \sqrt{\sin x} \, dx$$

$$= \int_0^{\pi/2} \overbrace{(\sin x)^{1/2}}^{u} \overbrace{\cos x \, dx}^{du}$$

❷ Determine the integral and transfer the limits of integration.

$$= \left[\frac{(\sin x)^{3/2}}{\frac{3}{2}} \right]_0^{\pi/2}$$

$$= \frac{2}{3} \left[\sqrt{(\sin x)^3} \right]_0^{\pi/2}$$

❸ Evaluate the last expression by plugging in the limits of integration.

$$= \frac{2}{3} \left[\sqrt{\left(\sin \frac{\pi}{2} \right)^3} - \sqrt{(\sin 0)^3} \right]$$

$$= \frac{2}{3} \left[\sqrt{1^3} - \sqrt{0^3} \right]$$

$$= \frac{2}{3}$$

Integrals of Exponential Functions

If u is a function of some other variable, say x, then $\int e^u\, du = e^u + c$ and $\int a^u\, du = \dfrac{a^u}{\ln a} + c$. **Another way to write the first integral above is** $\int e^{\text{some funct.}} \cdot (\text{der. of funct.}) = e^{\text{that funct.}} + c$. **For example,** $\int e^{\overset{u}{3x}} \cdot \overset{du}{3dx} = e^{\overset{u}{3x}} + c$ or just $e^{3x} + c$.

EXPONENTIAL INTEGRAL: EXAMPLE 1

$\int x e^{x^2}\, dx$

❶ Rewrite the integrand in a form that is closer to the exponential integral pattern.

$$\int x e^{x^2}\, dx$$
$$= \int e^{x^2} x\, dx$$
$$= \frac{1}{2} \int \underset{e^u}{e^{x^2}} \cdot \underset{du}{2x\, dx}$$

Notice how the $\dfrac{1}{2}$ and the 2 were used to get the correct du term.

❷ Use the exponential integral pattern to finish the problem.

$$= \frac{1}{2} e^{x^2} + c$$

EXPONENTIAL INTEGRAL: EXAMPLE 2

$\int \dfrac{e^{\ln x}}{x}\, dx$

❶ Rewrite the integrand to try to make the derivative of the exponent, $\ln x$, follow the term $e^{\ln x}$. Remember that $\dfrac{d}{dx}(\ln x) = \dfrac{1}{x}$.

$$\int \frac{e^{\ln x}}{x}\, dx$$
$$= \int \underset{u}{e^{\ln x}} \cdot \overset{du}{\frac{1}{x}\, dx}$$

❷ Apply the exponential integral formula.

$$= e^{\ln x} + c$$

....

EXPONENTIAL INTEGRAL: EXAMPLE 3 (WITH LIMITS OF INTEGRATION)

$$\int_1^9 \frac{e^{\sqrt{x}}}{\sqrt{x}}\,dx$$

❶ Rewrite the integrand so it follows the

$$\int e^{\text{some funct.}} \cdot (\text{der. of funct.}) = e^{\text{that funct.}} + c \ \text{pattern.}$$

$$\int_1^9 \frac{e^{\sqrt{x}}}{\sqrt{x}}\,dx$$

$$= \int_1^9 \frac{e^{x^{1/2}}}{x^{1/2}}\,dx$$

$$= \int_1^9 e^{x^{1/2}} \cdot x^{-1/2}\,dx$$

❷ You don't quite have the correct derivative of the exponent yet, but by inserting a ½ inside and a corresponding 2 outside, you'll get what you need.

$$= 2 \cdot \int_1^9 e^{\overset{u}{x^{1/2}}} \cdot \overset{\overbrace{}^{du}}{\tfrac{1}{2} x^{-1/2}}\,dx$$

❸ Complete the formula for the exponential integral and carry over the limits of integration.

$$= \frac{1}{2}\left[e^{\sqrt{x}}\right]_1^9$$

❹ Plug in the limits and simplify the result.

$$= \frac{1}{2}\left[e^{\sqrt{9}} - e^{\sqrt{1}}\right]$$

$$= \frac{1}{2}\left[e^3 - e\right]$$

Integrals of Exponential Functions (continued)

EXPONENTIAL INTEGRAL—BASE OTHER THAN *e*: EXAMPLE 1

$\int 2^{5x+7} dx$

1 You are trying to fit this into the $\int a^u du = \dfrac{a^u}{\ln a} + c$ formula. To do so, you need to have the derivative of the exponent $(5x + 7)$ follow the exponential term.

$$\int 2^{5x+7} dx$$

$$= \frac{1}{5} \cdot \int 2^{\overset{u}{\overbrace{5x+7}}} \cdot \overset{du}{\overbrace{5dx}}$$

2 Apply the formula for the non *e* base exponential integral.

$$= \frac{1}{5}\left[\frac{2^{5x+7}}{\ln 2}\right] + c$$

$$= \frac{2^{5x+7}}{5\ln 2} + c \text{ or } \frac{2^{5x+7}}{\ln 2^5} + c \text{ or } \frac{2^{5x+7}}{\ln 32} + c$$

EXPONENTIAL INTEGRAL—BASE OTHER THAN *e*: EXAMPLE 2

$\int x \cdot 5^{x^2} dx$

1 Try to fit the original integrand into the exponential integral pattern.

$$\int x \cdot 5^{x^2} dx$$

$$= \int 5^{x^2} \cdot x dx$$

$$= \frac{1}{2} \cdot \int 5^{\overset{u}{\overbrace{x^2}}} \cdot \overset{du}{\overbrace{2x dx}}$$

2 Complete the integral using the exponential integral formula.

$$= \frac{1}{2}\left[\frac{5^{\overset{u}{\overbrace{x^2}}}}{\underbrace{\ln 5}_{\ln \text{ of base}}}\right] + c$$

$$= \frac{5^{x^2}}{2\ln 5} + c \text{ or } \frac{5^{x^2}}{\ln 5^2} + c \text{ or } \frac{5^{x^2}}{\ln 25} + c$$

If u is a function of some other variable, say x, then $\int \frac{du}{u} = \ln|u| + c$. **Another way to write this is**

$$\int \frac{\text{der. of funct.}}{\text{funct.}} = \ln|\text{funct.}| + c.$$ **For example,** $\int \frac{3x^2}{x^3 - 5} dx = \int \frac{\overbrace{3x^2\, dx}^{du}}{\underbrace{x^3 - 5}_{u}} = \underbrace{\ln|x^3 - 5|}_{\ln|u|} + c.$

NATURAL LOG INTEGRAL: EXAMPLE 1

$$\int \frac{x + 1}{x^2 + 2x - 1} dx$$

❶ You want the top to be the derivative of the bottom. Since $\frac{d}{dx}(x^2 + 2x - 7) = 2x + 2$, you just need to multiply the numerator by 2 and then compensate with a $\frac{1}{2}$ outside the integral symbol.

$$\int \frac{x + 1}{x^2 + 2x - 1} dx$$

$$= \frac{1}{2} \cdot \int \frac{2(x + 1)}{x^2 + 2x - 1} dx$$

$$= \frac{1}{2} \cdot \int \frac{\overbrace{(2x + 2)\, dx}^{du}}{\underbrace{x^2 + 2x - 1}_{u}}$$

❷ Complete the formula, which results in $\ln|u| + c$.

$$= \frac{1}{2} \ln \left| \overset{u}{x^2 + 2x - 1} \right| + c$$

or just $\frac{1}{2} \ln|x^2 + 2x - 1| + c$

or $\ln\left(|x^2 + 2x - 1|\right)^{1/2} + c$

or $\ln\sqrt{x^2 + 2x - 1} + c$

Integrals That Result in a Natural Logarithmic Function (*continued*)

NATURAL LOG INTEGRAL: EXAMPLE 2 (USING A U-SUBSTITUTION)

$$\int \frac{e^{2x}+1}{e^{2x}+2x}\,dx$$

① Since you are trying to make this fit the $\int \frac{du}{u}$ pattern, let $u = e^{2x} + 2x$ and then proceed as shown at right.

let $u = e^{2x} + 2x$, then $du = (2e^{2x}+2)\,dx$

$$du = 2(e^{2x}+1)\,dx$$

or $\frac{1}{2}\,du = (e^{2x}+1)\,dx$

② Substitute the terms found in Step 1 into the appropriate spots in the original integrand.

Note: All expressions containing the variable x *have been replaced with* u *variable terms.*

$$\int \frac{e^{2x}+1}{e^{2x}+2x}\,dx$$

$$= \int \frac{(e^{2x}+1)\,dx}{e^{2x}+2x}$$

$$= \int \frac{\left(\frac{1}{2}\right)du}{u}$$

$$= \frac{1}{2} \cdot \int \frac{du}{u}$$

③ Now that the integral is in the $\int \frac{du}{u}$, complete the formula with the $\ln|u| + c$ portion.

$$= \frac{1}{2}\Big[\ln|u|\Big] + c$$

④ You need to get back to a solution in terms of x, not u, so substitute $u = e^{2x} + 2x$.

$$= \frac{1}{2}\ln\big|e^{2x}+2x\big| + c$$

or just $\frac{1}{2}\ln\big|e^{2x}+2x\big| + c$

NATURAL LOG INTEGRAL: EXAMPLE 3 (USING A U-SUBSTITUTION)

$$\int_2^3 \frac{x^3}{x^4+4}\,dx$$

1 Since you are trying to fit this to a natural log form $\int \frac{du}{u}$ let u be the denominator, so $u = x^4 + 4$.

let $u = x^4 + 4$, so that $du = 4x^3\,dx$

or $\frac{1}{4}\,du = x^3\,dx$

2 Since you are making a u-substitution, you may as well change the original x limits to the new u limits of integration.

$u = x^4 + 4$

for $x = 3 \rightarrow u = 3^4 + 4 = 85$

for $x = 2 \rightarrow u = 2^4 + 4 = 20$

3 Make a lot of substitutions, including the limits of integration, so that the original problem changes from x terms and limits to u terms and limits.

$$\int_2^3 \frac{x^3}{x^4+4}\,dx$$

$$= \int_2^3 \frac{x^3\,dx}{x^4+4} \quad \Leftarrow the \lim. of int. are x values$$

$$= \int_{20}^{85} \frac{\left(\frac{1}{4}\right)du}{u} \quad \Leftarrow the \lim. of int. are u values$$

$$= \frac{1}{4}\cdot\int_{20}^{85} \frac{du}{u}$$

4 Complete the integral, carry over the limits of integration, plug them in, and simplify the result.

$$= \frac{1}{4}\big[\ln|u|\big]_{20}^{85}$$

$$= \frac{1}{4}\big[\ln 85 - \ln 20\big]$$

$$= \frac{1}{4}\ln\left(\frac{85}{20}\right)$$

TIP

By changing from x limits to u limits of integration, you don't have to change back to x terms at the end.

or other forms, such as

$\frac{1}{4}\ln\left(\frac{17}{4}\right)$ or $\ln\left(\frac{17}{14}\right)^{1/4}$ or $\ln\sqrt[4]{\frac{17}{4}}$

Integrals of Trigonometric Functions

If u is a function of some variable, say x, then:

$$\int \cos u\, du = \sin u + c \qquad \int \sec u \tan u\, du = \sec u + c$$

$$\int \sin u\, du = -\cos u + c \qquad \int \csc^2 u\, du = -\cot u + c$$

$$\int \sec^2 u\, du = \tan u + c \qquad \int \csc u \cot u\, du = -\csc u + c$$

The integral formulas written above are just the result of reading backward the derivative formulas for the six trigonometric functions.

For example, since

$$\frac{d}{dx}(\sin u) = \cos u\, du \rightarrow \cos u\, du = \sin u + c$$

Similarly, since

$$\frac{d}{dx}(\sec u) = \sec u \tan u\, du \rightarrow \int \sec u \tan u\, du = \sec u + c$$

If u is a function of some variable, say x, then:

$$\int \tan u\, du = -\ln|\cos u| + c \;\; or \;\; \ln|\sec u| + c$$

$$\int \cot u\, du = \ln|\sin u| + c$$

$$\int \sec u\, du = \ln|\sec u + \tan u| + c$$

$$\int \csc u\, du = \ln|\csc u - \cot u| + c$$

The integral formulas above are a bit more difficult to verify. One example is shown at right.

$$\int \cot u\, du = \int \underbrace{\frac{\overbrace{\cos u}^{\text{der. of the funct.}}}{\underset{\text{the funct.}}{\sin u}}}\, du = \ln \left| \underset{\text{the funct.}}{\sin u} \right| + c$$

TRIGONOMETRIC INTEGRAL: EXAMPLE 1

$\int \cos 5x \, dx$

❶ To make this fit the $\int \cos u \, du$ pattern, you need the derivative of $5x$, namely 5, to follow the cos$5x$ term. So insert a 5 and compensate with a $\frac{1}{5}$ outside the integral symbol.

$\int \cos 5x \, dx$

$= \frac{1}{5} \cdot \int \left(\cos \underset{u}{5x} \right) \cdot \underset{du}{5dx}$

❷ You can now complete the integral using the right-hand side of the formula $\int \cos u \, du = \sin u + c$.

$= \frac{1}{5} \sin \underset{u}{5x} + c$

TRIGONOMETRIC INTEGRAL: EXAMPLE 2

$\int x^2 \sin(x^3) \, dx$

❶ Let's try the u-substitution method on this integral. But first, rewrite the integrand so it looks more like the $\int \sin u \, du$ pattern.

$\int x^2 \sin(x^3) \, dx$

$= \int \sin(x^3) \cdot x^2 \, dx$

❷ Let $u = x^3$ and then find du.

let $u = x^3$, then $du = 3x^2 \, dx$

or $\frac{1}{3} du = x^2 \, dx$

Integrals of Trigonometric Functions (continued)

❸ Substitute the expressions from Step 2 for the appropriate terms in Step 1.

$$= \int \sin\left(x^3\right) \cdot x^2 \, dx$$

$$= \int \sin u \cdot \frac{1}{3} \, du$$

$$= \frac{1}{3} \cdot \int \sin u \, du$$

❹ You can now complete the $\int \sin u \, du = -\cos u + c$ formula and replace the u with x^2 to finish the problem.

$$= \frac{1}{3}(-\cos u) + c$$

$$= -\frac{1}{3}\cos\left(x^3\right) + c$$

TRIGONOMETRIC INTEGRAL: EXAMPLE 3

$$\int \sec 3x \tan 3x \, dx$$

❶ This integral looks like the $\int \sec u \tan u \, du = \sec u + c$ form; you need a 3 to be the derivative of $3x$.

$$\int \sec 3x \tan 3x \, dx$$

$$= \frac{1}{3} \int \sec \underset{u}{\underline{3x}} \tan \underset{u}{\underline{3x}} \cdot \underset{du}{\underline{3dx}}$$

❷ Complete the right-hand side of the red formula above.

$$= \frac{1}{3}\sec \underset{u}{\underline{3x}} + c \text{ or just } \frac{1}{3}\sec 3x + c$$

TRIGONOMETRIC INTEGRAL: EXAMPLE 4

$$\int_0^{\pi/2} \frac{\sin x}{1 + \cos x} \, dx$$

❶ This looks like a natural log integral form, because the top is almost the derivative of the bottom. Let $u = 1 + \cos x$ and then find du. At the same time, let's change from the given x limits of integration to the new u limits.

with $u = 1 + \cos x$:

for $x = \dfrac{\pi}{2} \rightarrow u = 1 + \cos\left(\dfrac{\pi}{2}\right) = 1 + 0 = 1$

for $x = 0 \rightarrow u = 1 + \cos 0 = 1 + 1 = 2$

❷ Replace all x terms and limits with their appropriate u term counterparts.

$$\int_0^{\pi/2} \frac{\sin x}{1 + \cos x} \, dx$$

$$= \int_0^{\pi/2} \frac{\sin x \, dx}{1 + \cos x} \Leftarrow x \text{ terms and } x \text{ lim. of int.}$$

$$= \int_2^1 \frac{-du}{u} \Leftarrow u \text{ terms and } u \text{ lim. of int.}$$

$$= -\int_2^1 \frac{du}{u}$$

❸ Notice that the upper limit of integration is smaller than the lower limit. Since you are probably used to having the upper limit bigger than the lower limit, switch the limits and also take the opposite of the integral.

$$= -\left[-\int_1^2 \frac{du}{u} \right]$$

$$= \int_1^2 \frac{du}{u}$$

❹ Complete the natural log integral form with the right-hand side of the formula $\int \frac{du}{u} = \ln|u|$. (The "$+ c$" was dropped because there are limits of integration involved in this problem.)

$$= \Big[\ln|u| \Big]_1^2$$

❺ Plug in the limits and simplify the result.

$$= \ln|2| - \ln|1|$$

$$= \ln 2 - \ln 1$$

$$= \ln 2 - 0$$

$$= \ln 2$$

Integrals of Trigonometric Functions *(continued)*

TRIGONOMETRIC INTEGRAL: EXAMPLE 5

$\int \frac{\cos^4 x}{\csc x}\, dx$

1 In searching for an appropriate integration technique to use here, the one that first comes to mind is a natural log. But in this case, the top is not the derivative of the bottom, so the natural log won't work. Let's try rewriting the integrand to see if some other technique presents itself.

$$\int \frac{\cos^4 x}{\csc x}\, dx$$
$$= \int \cos^4 x \cdot \frac{1}{\csc x}\, dx$$
$$= \int \cos^4 x \cdot \sin x\, dx$$
$$= \int (\cos x)^4 \cdot \sin x\, dx$$

2 Now it looks like a General Power Rule pattern, but the derivative of the inside function, cos *x*, is actually –sin *x*, so we need a negative sign to get the General Power Rule just right.

$$= -\int \overbrace{(\cos x)}^{u}{}^{\overset{n}{4}} \cdot \overbrace{(-\sin x)}^{du}\, dx$$

3 Complete the General Power Rule formula: $\int u^n\, du = \frac{u^{n+1}}{n+1} + c$

$$= -\frac{\overbrace{(\cos x)}^{u}{}^{\overset{n+1}{5}}}{\underbrace{5}_{n+1}} + c$$
$$= -\cos^5 x + c$$

TRIGONOMETRIC INTEGRAL: EXAMPLE 6

$\int \sec^4 x \tan x\, dx$

1 Let's use a bit of an unusual strategy on this integral. First, rewrite the integrand, making the power of sec x more obvious.

$$\int \sec^4 x \tan x\, dx$$
$$= \int (\sec x)^4 \cdot \tan x\, dx$$

2 This almost looks like a General Power Rule pattern, except that the derivative of sec x is sec x tan x. So "borrow" a sec x *term from the* $(\sec x)^4$.

$$= \int (\sec x)^3 \cdot \sec x \tan x\, dx$$
$$= \int \underbrace{(\sec x)^3}_{u} \cdot \underbrace{\sec x \tan x\, dx}_{du}$$

3 Now that your integral fits the $\int u^n\, du = \dfrac{u^{n+1}}{n+1} + c$ pattern, complete the right-hand side of the General Power Rule formula.

$$= \frac{\overbrace{(\sec x)}^{u}{}^{\overbrace{4}^{n+1}}}{\underbrace{4}_{n+1}} + c$$
$$= \frac{1}{4} \sec^4 x + c$$

Integrals That Result in an Inverse Trigonometric Function

If u is a function of some variable, say x, and a is some constant, then

$$\int \frac{du}{\sqrt{a^2 - u^2}} = \arcsin \frac{u}{a} + c$$

$$\int \frac{du}{a^2 + u^2} = \frac{1}{a} \arctan \frac{u}{a} + c$$

$$\int \frac{du}{u\sqrt{u^2 - a^2}} = \frac{1}{a} arc \sec \frac{|u|}{a} + c$$

At right are brief examples of each of these integral formulas.

$$\int \frac{3dx}{\sqrt{25 - 9x^2}} = \int \frac{\overbrace{3dx}^{du}}{\sqrt{\underbrace{5^2}_{a} - \underbrace{(3x)^2}_{u}}} = \arcsin \frac{\overbrace{3x}^{u}}{\underbrace{5}_{a}} + c \text{ or } \arcsin \frac{3x}{5} + c$$

$$\int \frac{10xdx}{49 + 25x^4} = \int \frac{\overbrace{10xdx}^{du}}{\underbrace{7^2}_{a} + \underbrace{(5x^2)^2}_{u}} = \frac{1}{7} \arctan \frac{\overbrace{5x^2}^{u}}{\underbrace{7}_{a}} + c \text{ or } \frac{1}{7} \arctan \frac{5x^2}{7} + c$$

$$\int \frac{5dx}{5x\sqrt{25x^2 - 16}} = \int \frac{\overbrace{5dx}^{du}}{\underbrace{5x}_{u}\sqrt{\underbrace{(5x)^2}_{u} - \underbrace{4^2}_{a}}} = \frac{1}{4} arc \sec \frac{\overbrace{|5x|}^{u}}{\underbrace{4}_{a}} + c = \frac{1}{4} arc \sec \frac{|5x|}{4} + c$$

. . . .

INVERSE TRIGONOMETRIC INTEGRAL: EXAMPLE 1

$$\int \frac{dx}{\sqrt{25 - 4x^2}}$$

❶ How do you determine which inverse trigonometric form to use? In this case, the denominator has a radical that contains a constant minus a function square, as in $\sqrt{a^2 - u^2}$. So this is an arcsin form. You need to write the integrand in that form.

$$\int \frac{dx}{\sqrt{25 - 4x^2}}$$

$$= \int \frac{dx}{\sqrt{\underbrace{5^2}_{a} - \underbrace{(2x)^2}_{u}}}$$

❷ You're close to the correct form, except that the numerator is not the derivative of the u term, $2x$. Since $\frac{d}{dx}(2x) = 2$, you need to insert a 2 on top of the integrand and compensate with a $\frac{1}{2}$ outside.

$$= \frac{1}{2} \int \frac{\overbrace{2dx}^{du}}{\sqrt{\underbrace{5^2}_{a} - \underbrace{(2x)^2}_{u}}}$$

❸ Now the integral fits the left side of the formula $\int \frac{du}{\sqrt{a^2 - u^2}} = \arcsin \frac{u}{a} + c$, so just complete the right-hand side.

$$= \frac{1}{2}\left[\arcsin \frac{\overset{u}{\overbrace{2x}}}{\underset{a}{\underbrace{5}}} \right] + c \text{ or } \frac{1}{2} \arcsin \frac{2x}{5} + c$$

INVERSE TRIGONOMETRIC INTEGRAL: EXAMPLE 2

$$\int \frac{x^2 \, dx}{16 + x^6}$$

❶ If this is an inverse trigonometric integral, it has to be arctan, because there is no radical in the denominator. Rewrite the integrand to try to get it into the arctan pattern.

$$\int \frac{x^2 \, dx}{16 + x^6}$$
$$= \int \frac{x^2 \, dx}{4^2 + \left(x^3\right)^2}$$

❷ Let's use the u-substitution this time—it appears that $u = x^3$. Find du also.

let $u = x^2$, then $du = 3x^2 \, dx$

or $\frac{du}{3} = x^2 \, dx$

3 Substitute the u terms from Step 2 for the appropriate x terms in Step 1.

$$= \int \frac{x^2\, dx}{4^2 + \left(x^3\right)^2}$$

$$= \int \frac{\frac{du}{3}}{4^2 + u^2}$$

$$= \frac{1}{3} \int \frac{du}{4^2 + u^2}$$

4 With $a = 4$, this now fits the arctan formula:

$\int \frac{du}{a^2 + u^2} = \frac{1}{a} \arctan \frac{u}{a} + c$. Now complete the right-hand side of the formula.

$$= \frac{1}{3}\left[\frac{1}{4} \arctan \frac{u}{4}\right] + c$$

$$= \frac{1}{12} \arctan \frac{u}{4} + c$$

5 Last, make the substitution of $u = x^3$ to complete the problem by returning to x terms.

$$= \frac{1}{12} \arctan \frac{x^3}{4} + c \text{ or } \frac{1}{12} \arctan \frac{x^3}{4} + c$$

You will occasionally encounter an integral that by itself cannot be integrated. In some of these cases you will have to first alter the form of the integrand in order to use multiple techniques to complete the integration process.

"Combo" Technique: General Power Rule and an Arcsin

$$\int \frac{x+4}{\sqrt{9-x^2}}\,dx$$

❶ This does not fit any of the forms you have studied so far. Split the original integral into two separate integrals to see if that helps.

$$\int \frac{x+4}{\sqrt{9-x^2}}\,dx$$

$$= \int \left(\frac{x}{\sqrt{9-x^2}} + \frac{4}{\sqrt{9-x^2}} \right) dx$$

$$= \int \frac{x}{\sqrt{9-x^2}}\,dx + \int \frac{4}{\sqrt{9-x^2}}\,dx$$

❷ The blue integral is in a General Power Rule form, and the red integral is in an arcsin form. Modifying the integrands further will make these forms more apparent.

$$= \int (9-x^2)^{-1/2}\cdot x\,dx + 4\cdot \int \frac{dx}{\sqrt{3^2-(x)^2}}$$

❸ The first integral needs a -2 inside (with a $-\frac{1}{2}$ outside); the second integral's form is fine.

$$= -\frac{1}{2}\cdot \int (9-x^2)^{-1/2}\cdot(-2)x\,dx + 4\cdot \int \frac{dx}{\sqrt{3^2-(x)^2}}$$

❹ Use the General Power Rule for the blue and an arcsin for the red integral.

$$= -\frac{1}{2}\left[(9-x^2)^{1/2}\right] + 4\cdot\left[\arcsin \frac{x}{3}\right] + c$$

$$= -\frac{1}{2}\sqrt{9-x^2} + 4\arcsin \frac{x}{3} + c$$

Combinations of Functions and Techniques *(continued)*

Regarding "Look-Alike" Integrals

The integral forms in each colored pair listed at right are frequently mistaken for one another.

$$\int \frac{x^3}{9 + x^4}\, dx \text{ and } \int \frac{x}{9 + x^4}\, dx \qquad \int \frac{x^3}{\sqrt{9 + x^4}}\, dx \text{ and } \int \frac{x}{\sqrt{9 - x^4}}\, dx$$

---

❶ Take a closer look at the red pair as you rewrite their integrands to reveal the particular technique appropriate to that integral.

$$\int \frac{x^3}{9 + x^4}\, dx$$

$$= \frac{1}{4} \cdot \underbrace{\int \frac{4x^3}{9 + x^4}}_{\text{natural log}} dx$$

$$\int \frac{x}{9 + x^4}\, dx$$

$$= \int \frac{x}{9 + \left(x^2\right)^2}\, dx$$

$$= \frac{1}{2} \cdot \underbrace{\int \frac{2x}{9 + \left(x^2\right)^2}\, dx}_{\text{arctan}}$$

---

❷ Take a closer look at the blue pair as you rewrite their integrands to reveal the particular technique appropriate to that integral.

$$\int \frac{x^3}{\sqrt{9 + x^4}}\, dx$$

$$= \int \left(9 + x^4\right)^{-1/2} x^3\, dx$$

$$= \frac{1}{4} \cdot \underbrace{\int \left(9 + x^4\right)^{-1/2} \cdot 4x^3\, dx}_{\text{general power rule}}$$

$$\int \frac{x}{\sqrt{9 - x^4}}\, dx$$

$$= \int \frac{x}{\sqrt{9 - \left(x^2\right)^2}}\, dx$$

$$= \frac{1}{2} \cdot \underbrace{\int \frac{2x}{\sqrt{9 - \left(x^2\right)^2}}\, dx}_{\text{arcsin}}$$

Sometimes you encounter an integral that doesn't seem to fit any of the more common integration forms. In this case, the u-substitution technique may be useful.

EXAMPLE 1

$\int x \sqrt{x-1}\, dx$

❶ Let u be the most complicated part of the integrand; in this case, let $u = \sqrt{x-1}$. Now find dx and solve for x in terms of u.

let $u = \sqrt{x-1}$, or in another form $u = (x-1)^{1/2}$ to be used later

then $u^2 = x - 1$

so that $u^2 + 1 = x$

and $2u\, du = dx$

---

❷ Now you can make a large series of substitutions, plugging data from Step 2 into the original integral.

$\int x \sqrt{x-1}\, dx$

$= \int (u^2 + 1) \cdot u \cdot 2u\, du$

---

❸ Expand and simplify the integrand in Step 2.

$= \int (u^2 + 1) \cdot 2u^2\, du$

$= 2 \cdot \int (u^4 + u^2)\, du$

---

❹ Use the Simple Power Rule on both terms of the integrand.

$= 2 \cdot \left[\dfrac{u^5}{5} + \dfrac{u^3}{3} \right] + c$

$= \dfrac{2}{5} u^5 + \dfrac{2}{3} u^3 + c$

⑤ You need to replace the u terms with their appropriate x terms, this time using the substitution $u = (x - 1)^{1/2}$ found in Step 1.

$$= \frac{2}{5}\left((x-1)^{1/2}\right)^5 + \frac{2}{3}\left((x-1)^{1/2}\right)^3 + c$$

$$= \frac{2}{5}(x-1)^{5/2} + \frac{2}{3}(x-1)^{3/2} + c$$

or in another form

$$= \frac{6}{15}(x-1)^{5/2} + \frac{10}{15}(x-1)^{3/2} + c$$

$$= \frac{2}{15}(x-1)^{3/2}\left[3(x-1)+5\right]$$

$$= \frac{2}{15}(x-1)^{3/2}(3x+2)$$

......

EXAMPLE 2

$$\int_{-2}^{1} \frac{x}{\sqrt{x+3}}\, dx$$

① Let's work Example 2, but this time change the x limits to u limits. Letting $u = \sqrt{x+3}$, proceed as before in finding dx and x in terms of u.

letting $u = \sqrt{x+3}$

then $u^2 = x + 3$

so that $u^2 - 3 = x$

and then $2u\, du = dx$

for $x = 1 : u = \sqrt{1+3} = 2$

for $x = -2 : u = \sqrt{-2+3} = 1$

2 Make your long list of substitutions into the original integral, changing all x terms and limits to u terms and limits.

$$\int_{-2}^{1} \frac{x}{\sqrt{x+3}} \, dx \Leftarrow \text{ this has } x \text{ lim. of int.}$$

$$= \int_{1}^{2} \frac{(u^2-3)}{\cancel{u}} \cdot 2\cancel{u} \, du \Leftarrow u \text{ terms and } u \text{ lim. of int.}$$

$$= 2\int_{1}^{2} (u^2-3) \, du$$

3 Integrate and then plug in the u limits.

$$= \left[2 \cdot \frac{u^3}{3} - 6u \right]_{1}^{2}$$

$$= \left[\frac{2}{3} u^3 - 6u \right]_{1}^{2}$$

$$= \left[\frac{2}{3} \cdot 2^3 - 6 \cdot 2 \right] - \left[\frac{2}{3} \cdot 1^3 - 6 \cdot 1^3 \right]$$

$$= \left[\frac{16}{3} - 12 \right] - \left[\frac{2}{3} - 6 \right]$$

$$= \frac{16}{3} - 12 - \frac{2}{3} + 6$$

$$= -\frac{4}{3}$$

TIP

If we had not changed from the x limits to the u limits of integration the computation at the end of the problem would have been much different — yet the final answer would be the same. Try it and see what happens.

Solving Variables Separable Differential Equations

When you find the derivative of some function, the resulting equation is also known as a **differential equation.** For example, $\frac{dy}{dx} = xy + 7x$, or $y' = \frac{x^2}{y}$, or $(3 + x^3)y' - 2xy = 0$. Our goal in this section is to determine from which original equation, or function, the differential equation was derived.

GENERAL SOLUTION 1

Using this technique, you will separate the variables (hence the name for the technique) so that all the y and dy terms are on one side of the equation and all the x and dx terms are on the other. Then, by integrating both sides, you will arrive at an equation involving just x and y terms.

Solve the differential equation $\frac{dy}{dx} = \frac{x^2}{y}$ and write the solution in the form ". . . = some constant."

❶ Cross-multiply to get the dy and y terms on the left side and the dx and x terms on the right.

$$\frac{dy}{dx} = \frac{x^2}{y}$$

$$y \, dy = x^2 \, dx$$

❷ Now that you have separated the variables, integrate both sides of the equation.

Notice that is only one "+ c." If you were to put "+ d" on the left and "+ m" on the right, they would eventually combine to make some third constant "+ c."

$$\int y \, dy = \int x^2 \, dx$$

$$\frac{y^2}{2} = \frac{x^3}{3} + c$$

❸ Put all the y and x terms on the left (the answer form that was requested).

$$\frac{y^2}{2} - \frac{x^3}{3} = c \text{ or just } \frac{y^2}{2} - \frac{x^3}{3} = c$$

❹ If you chose to, or if it was requested of you, you could eliminate the fractions by multiplying all terms of the equation by 6.

$m = 6c$ is just another constant.

$$\frac{y^2}{2} - \frac{x^3}{3} = c$$

$$6 \cdot \frac{y^2}{2} - 6 \cdot \frac{x^3}{3} = 6 \cdot c$$

$$3y^2 - 2x^3 = 6c$$

$$3y^2 - 2x^3 = m$$

....

GENERAL SOLUTION 2

For the differential equation $\dfrac{dy}{dx} = xy + 3x$, write its solution in the form "$y = \ldots$"

❶ This problem takes a little more creativity to get the variables separated.

$$\frac{dy}{dx} = xy + 3x$$

$$\frac{dy}{dx} = x(y + 3)$$

❷ Multiply both sides by dx.

$$dy = x(y + 3)dx$$

❸ Divide both sides by $y + 3$, and the variables will finally be separated.

$$\frac{dy}{y + 3} = x \, dx$$

❹ You're ready to find the integral of each side.

$$\int \frac{dy}{y + 3} = \int x \, dx$$

$$\ln|y + 3| = \frac{x^2}{2} + c$$

⑤ Rewrite the last natural log equation as an exponential one instead, and then simplify the right-hand side.

$$\left| y + 3 \right| = e^{x^2 \div 2 + c}$$

$$\left| y + 3 \right| = e^{x^2 \div 2} \cdot e^c$$

TIP

If $\ln y = x$, then $y = e^x$.

$$\Downarrow$$

$$\left| y + 3 \right| = e^{x^2 \div 2} \cdot m$$

$$\left| y + 3 \right| = m \cdot e^{x^2 \div 2}$$

⑥ Get rid of the absolute value symbol on the left and then simplify the constant on the right.

$$y + 3 = \pm m \cdot e^{x^2 \div 2}$$

$$y = k e^{x^2 \div 2} - 3$$

GENERAL SOLUTION 3

Solve the differential equation $(3 + x2)y' - 2xy = 0$ and write the solution in the form "$y = \ldots$"

① Add $2xy$ to both sides and then replace the y' with $\dfrac{dy}{dx}$

$$\left(3 + x^2 \right) y' - 2xy = 0$$

$$\left(3 + x^2 \right) y' = 2xy$$

$$\left(3 + x^2 \right) \frac{dy}{dx} = 2xy$$

$$\left(3 + x^2 \right) \frac{dy}{dx} = 2xy$$

② Multiply both sides by dx.

$$(3 + x^2) dy = 2xy \, dx$$

③ Divide both sides by y.

$$\left(3 + x^2 \right) \frac{dy}{y} = 2x \, dx$$

4 Divide both sides of the equation by $3 + x^2$.

$$\frac{dy}{y} = \frac{2x}{3 + x^2} dx$$

5 With the variables separated, you can integrate both sides.

$$\int \frac{dy}{y} = \int \frac{2x}{3 + x^2} dx$$

$$\ln|y| = \ln|3 + x^2| + c$$

$$\ln|y| = \ln(3 + x^2) + c \quad \text{Note that } 3 + x^2 \text{ is always positive.}$$

6 There are two ways to deal with this double natural log situation. We'll use one method here and then demonstrate the other method later in the problem.

Replace the c with $\ln e^c$.

$$\ln|y| = \ln(3 + x^2) + \ln e^c$$

7 Use your log properties to rewrite the right-hand side of the equation as a single natural log.

$$\ln|y| = \ln\left[(3 + x^2) \cdot e^c\right]$$

8 Since the natural log of left quantity equals the natural log of the right quantity, you can get rid of the *ln* on both sides.

$$|y| = (3 + x^2) \cdot e^c$$

$$\Downarrow$$

> **TIP**
>
> e^c is just another constant, say m.

$$|y| = (3 + x^2) \cdot m$$

$$|y| = m(3 + x^2)$$

9 Take out the absolute value symbols on the left and insert a $\pm$ sign on the right.

$$y = \pm m(3 + x^2)$$

$$\Downarrow$$

> **TIP**
>
> $\pm m$ is just another constant, say k.

$$y = k(3 + x^2)$$

10 Here's the promised alternate solution, beginning with Step 5.

$$\int \frac{dy}{y} = \int \frac{2x}{3 + x^2} \, dx$$

$$\ln|y| = \ln|3 + x^2| + c$$

$$\ln|y| = \ln(3 + x^2) + c \quad \text{Note that } 3 + x^2 \text{ is always positive.}$$

11 If $m = r$, then $e^m = e^r$ must also be true. So in our case, $e^{\text{left side of equation}} = e^{\text{right side of equation}}$.

$$e^{\ln|y|} = e^{\ln(3+x^2)+c}$$

$$e^{\ln|y|} = e^{\ln(3+x^2)} \cdot e^c$$

$$\Downarrow$$

$$e^{\ln|y|} = e^{\ln(3+x^2)} \cdot m$$

$$e^{\ln|y|} = m \cdot e^{\ln(3+x^2)}$$

12 Using the natural log property $e^{\ln x} = x$, simplify both sides of the equation.

$$y = m(3 + x^2) \text{ or just } y = m(3 + x^2)$$

Particular Solution

The general solution of a differential equation will always have some constant in the solution. If you are given additional information, often called an **initial condition**, you will be able to find a particular value of the constant, and thus a particular solution to the differential equation.

Given the initial condition of $y(1) = 3$, find the particular solution of the differential equation $yy' - 3x = 0$.

1 Replace the y' with $\frac{dy}{dx}$.

$$y \frac{dy}{dx} - 3x = 0$$

❷ Add $3x$ to both sides.

$$y\frac{dy}{dx} = 3x$$

❸ Multiply both sides by dx to separate your variables.

$$y\,dy = 3x\,dx$$

❹ Integrate both sides.

$$\int y\,dy = \int 3x\,dx$$
$$\int y\,dy = 3\cdot\int x\,dx$$
$$\frac{y^2}{2} = 3\cdot\frac{x^2}{2} + c$$

❺ Multiply both sides by 2 and simplify the new constant on the right side.

$$y^2 = 3x^2 + 2c$$
$$\Downarrow$$
$$y^2 = 3x^2 + m$$

You just found the general solution.

❻ Using the initial condition that $y(1) = 3$ (that is, when $x = 1$, then $y = 3$), you can find a particular value of the constant m and thus a particular solution to the differential equation.

$$y^2 = 3x^2 + m$$
$$(3)^2 = 3(1)^2 + m$$
$$-6 = m$$

Therefore, the particular solution is
$$y^2 = 3x^2 - 6.$$

Applications of Integration

The chapter opens with integration of functions related to the motion of an object and then moves on to finding the area of a region bounded by the graphs of two or more functions using an appropriate integral.

Revolving a bounded region about a given vertical or horizontal line produces a solid the volume of which you will be able to compute using an appropriate integral. Three methods for doing this are introduced: disk, washer, and shell.

Acceleration, Velocity, and
Position 247

Area between Curves:
Using Integration 250

Volume of Solid of Revolution:
Disk Method 260

Volume of Solid of Revolution:
Washer Method 268

Volume of Solid of Revolution:
Shell Method. 275

In Chapters 3–6 on derivatives, you found a way to move from the position to the velocity and then on to the acceleration function by differentiating. Now you will reverse the process, going from acceleration back to the position function by integrating.

If acceleration function is $a(t)$, then velocity function is $v(t) = \int a(t)\, dt$, and position function is $s(t) = \int v(t)\, dt$.

MOTION PROBLEM: ACCELERATION TO VELOCITY

The acceleration function, $a(t)$, for an object is given by $a(t) = 36t - 168t + 120$. If $v(1) = 28$, find the velocity function, $v(t)$.

❶ Beginning with the acceleration function, find its integral to get the velocity function.

$$a(t) = 36t^2 - 168t + 120$$
$$v(t) = \int a(t)\, dt = \int (36t^2 - 168t + 120)\, dt$$
$$v(t) = \frac{36t^3}{3} - \frac{168t^2}{2} + 120t + c$$
$$v(t) = 12t^3 - 84t^2 + 120t + c$$

❷ Using the given data that $v(1) = 28$, substitute 1 for t and 28 for $v(1)$ to solve for the constant c.

$$v(1) = 12(1)^3 - 84(1)^2 + 120(1) + c$$
$$\Downarrow$$
$$28 = 12 - 84 + 120 + c$$
$$-20 = c$$

❸ Replace the c with -20 in the velocity function.

$$v(t) = 12t^3 - 84t^2 + 120t + c$$
$$\Downarrow$$
$$v(t) = 12t^3 - 84t^2 + 120t - 20$$

MOTION PROBLEM: VELOCITY TO POSITION

If $s(1) = 38$ and $v(t) = 12t - 84t + 120t + 20$, find the position function, $s(t)$.

❶ Integrate $v(t)$ to get $s(t)$.

$$s(t) = \int v(t)\,dt = \int \left(12t^3 - 84t^2 + 120t + 20\right) dt$$

$$s(t) = \frac{12t^4}{4} - \frac{84t^3}{3} + \frac{120t^2}{2} + 20t + k$$

$$s(t) = 3t^4 - 28t^3 + 60t^2 + k$$

Note: Since c is used for the constant in the problem above, k is used here so that there is no confusion about which constant goes with which problem.

❷ Use the data $s(1) = 38$ to make appropriate substitutions and then solve for k.

$$s(1) = 3(1)^4 - 28(1)^3 + 60(1)^2 + k$$

$$\Downarrow$$

$$38 = 3 - 28 + 60 + k$$

$$3 = k$$

❸ Replace the k with 3 in the velocity function.

$$s(t) = 3t^4 - 28t^3 + 60t^2 + k$$

$$\Downarrow$$

$$s(t) = 3t^4 - 28t^3 + 60t^2 + 3$$

Motion Problem: Acceleration to Position

If the acceleration function of an object is given by $a(t) = \sin t + \cos t$ and $v(\pi) = 2$, while $s(\pi) = 1$, find the position function $s(t)$.

1 Find the velocity function $v(t)$ by integrating the acceleration function $a(t)$, and then find the position function $s(t)$ by integrating the velocity function $v(t)$.

$$v(t) = \int a(t)\, dt = \int (\sin t + \cos t)\, dt$$
$$v(t) = -\cos t + \sin t + c$$

2 Use the data $v(\pi) = 2$ to find the value of c.

$$v(\pi) = -\cos \alpha + \sin \pi + c$$
$$\Downarrow$$
$$2 = -(-1) + 0 + c$$
$$1 = c$$
therefore $v(t) = -\cos t + \sin t + 1$

3 Find the position function $s(t)$.

$$s(t) = \int v(t)\, dt = \int (-\cos t + \sin t + 1)\, dt$$
$$s(t) = -\sin t \quad \cos t \quad t + k$$

4 You are given additional data, $s(\pi) = 1$, which will allow you to find the value of k.

$$s(\pi) = -\sin \pi - \cos \pi + \pi + k$$
$$\Downarrow$$
$$1 = -(0) - (-1) + \pi + k$$
$$-\pi = k$$
thus $s(t) = -\sin t - \cot t + t - \pi$

Area between Curves: Using Integration

There are two scenarios to consider when finding the area of the region bounded by the graphs of two or more equations: Either the graphs of the equations do not intersect or the graphs of the equations intersect at one or more points. You will use integration to compute the area of the bounded region.

Scenario 1: The Graphs of the Equations Do Not Intersect

❶ The region, the area of which you are trying to compute, is bounded by the graphs of two functions that do not intersect $y_1 = f(x)$ and $y_2 = g(x)$ and the graphs of two vertical or horizontal lines ($x = a$ *and* $x = b$).

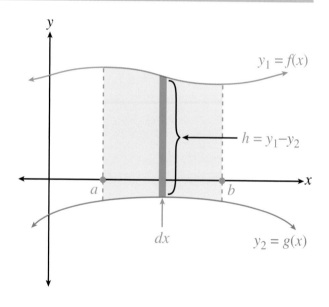

❷ Notice the thin green "representative" rectangle. Its base is a small change in x, typically labeled as Δx, or just dx. Its height is the difference in y coordinates for the two functions—in this case $y_1 - y_2$.

area of "rep." rect. = height · base = $(y_1 - y_2)dx$

❸ If you were to add the areas of an infinite number of very thin such rectangles, you could find their sum by using just an integral.

$$Area = \int_a^b (y_1 - y_2)\, dx$$

❹ If you substitute $y_1 = f(x)$ and $y_2 = g(x)$, you end up with an integral representing the area of the region bounded by the graphs of the two given functions and the two given vertical lines.

$$Area = \int_a^b \left(f(x) - g(x) \right) dx$$

Notice that $f(x)$ is the top function and $g(x)$ is the bottom function in the diagram shown in Step 1. (See p. 250.)

Scenario 2: The Graphs of the Two Functions Intersect One or More Times

❶ The region, the area of which you are trying to compute, is bounded by the graphs of two functions $y_1 = f(x)$ and $y_2 = g(x)$, which intersect at points where $x = a$, $x = b$, *and* $x = c$.

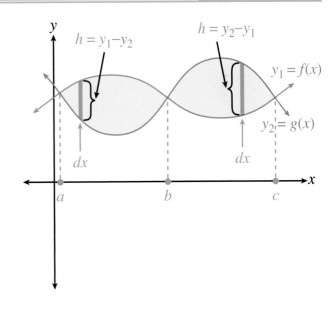

❷ Each green shaded region has a thin green rectangle with base dx, but they have different heights. The one on the left has a height of $y_1 - y_2$, and the one on the right has a height of $y_2 - y_1$; notice that in either case, the height is just the top function minus the bottom function.

area of left rep. rect. $= (y_1 - y_2)dx$
area of right rep. rect. $= (y_2 - y_1)dx$

❸ If you were to add the areas of an infinite number of very thin rectangles in each region, their sum could be found by the sum of two integrals.

total area = area of left region + area of right region

$$= \int_a^b (y_1 - y_2)\,dx + \int_b^c (y_2 - y_1)\,dx$$

❹ Make the substitutions $y_1 = f(x)$ *and* $y_2 = g(x)$ to get the area in terms of the given functions and the x coordinates of the points of intersection of their graphs.

$$total\ area = \int_a^b \left(f(x) - g(x) \right) dx + \int_b^c \left(g(x) - f(x) \right) dx$$

EXAMPLE 1

Find the area of the region bounded by the graphs of $y = x$, $y = 0$, $x = 1$, and $x = 4$.

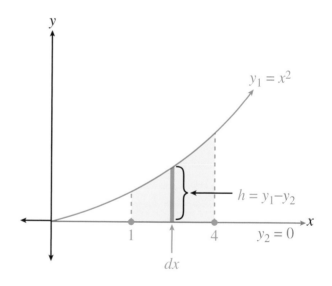

❶ Find the area of the green representative rectangle.

area rep. rect. $= hb = (y_1 - y_2)dx$

❷ If you find the sum of lots of these rectangles, you get an integral that represents the area of the green shaded region.

$$A = \int_1^4 \left(y_1 - y_x\right) dx$$

❸ Since the term at the end of the integral is dx, the integrand can contain x variables and/or numbers. Substitute $y_1 = x^2$ and $y_2 = 0$ and then simplify.

$$A = \int_1^f \left(x^2 - 0\right) dx$$

$$A = \int_1^4 x^2 \, dx$$

❹ Evaluate the integral to get the area of the green shaded region.

$$A = \left[\frac{x^3}{3}\right]_1^4 = \frac{64}{3} - \frac{1}{3} = \frac{63}{3}$$

$$A = 21$$

EXAMPLE 2

Find the area of the region bounded by the graphs of $y = 3x - x^2$ *and* $y = 0$.

❶ You have not been given the limits of integration, so you will have to find them by setting the equations equal to one another and then solving for x.

$$3x - x^2 = 0$$
$$x(3 - x) = 0$$
$$x = 0, x = 3$$

❷ Sketch a graph showing the bounded region and a representative rectangle indicating its base and its height.

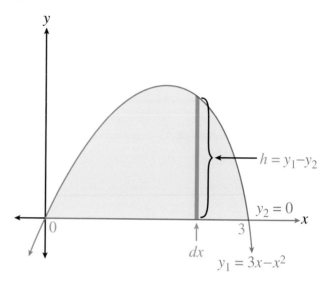

❸ Write the formula to find the area of the representative rectangle.

area rep. rect. $= hb$

$$= (y_1 - y_2)dx$$

❹ Set up the integral that represents the sum of lots of these rectangles, and thus the area of the green shaded region.

$$A = \int_0^3 (y_1 - y_2)\, dx$$

❺ Substitute the appropriate x or number equivalents for the terms y_1 *and* y_2 and simplify (since there is a dx term at the end, you need all x or numerical terms in the integrand) $y_1 = 3x - x^2$ and $y_2 = 0$.

$$A = \int_0^3 \left((3x - x^2) - 0\right) dx$$

$$A = \int_0^3 (3x - x^2)\, dx$$

❻ Evaluate the integral, finding the area of the desired bounded region.

$$A = \left[\frac{3x^2}{2} - \frac{x^3}{3}\right]_0^3 = \frac{27}{2} - 9$$

$$A = \frac{9}{2}$$

EXAMPLE 3

Find the area of the region bounded by the graphs of $y = x^3 - 9x$ and $y = 0$.

❶ Find the x coordinates of any points of intersection of the graphs of the two functions. This will also aid in sketching the graph of the first function.

$$x^3 - 9x = 0$$
$$x(x^2 - 9x) = 0$$
$$x(x + 3)(x - 3) = 0$$
$$x = 0,\ x = -3,\ x = 3$$

Note: These will serve as your limits of integration.

❷ Sketch the graphs of the two functions, indicating the bounded regions whose area you are going to compute. Also, show a representative rectangle for each region, along with its base and appropriate height.

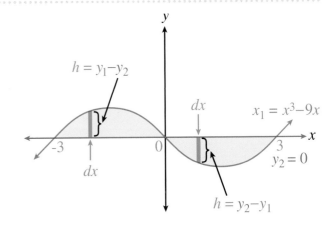

❸ Find the area of the representative rectangle for each region and then find their sum.

area of left rep. rect. $= (y_1 - y_2)dx$
area of right rep. rect. $= (y_2 - y_1)dx$

❹ Set the sum of the two integrals: one to find the area of the green shaded region from $x = -3$ to $x = 0$, and the other from $x = 0$ to x to 3.

$$total\ area = \int_{-3}^{0} (y_1 - y_2)\, dx + \int_{0}^{3} (y_2 - y_1)\, dx$$

Notice that for each integrand, it's the top function minus the bottom function within each interval of integration.

5 Substitute $y_1 = x^3 - 9x$ *and* $y_2 = 0$ and then simplify each integrand.

Now the integrands will match variables with the dx term at the end of each integral. All expressions will be in terms of the variable x.

$$total\ area = \int_{-3}^{0} \left((x^3 - 9x) - 0 \right) dx + \int_{0}^{3} \left(0 - (x^3 - 9x) \right) dx$$

$$= \int_{-3}^{0} (x^3 - 9x)\, dx + \int_{0}^{3} (-x^3 + 9x)\, dx$$

6 Evaluate each integral and find their sum. This will be the sum of the areas of the two green shaded regions.

$$total\ area = \int_{-3}^{0} (x^3 - 9x)\, dx + \int_{0}^{3} (-x^3 + 9x)\, dx$$

$$= \left[\frac{x^4}{4} - \frac{9x^2}{2} \right]_{-3}^{0} + \left[-\frac{x^4}{4} + \frac{9x^2}{2} \right]_{0}^{3}$$

$$= \left[0 - \left(\frac{81}{4} - \frac{81}{2} \right) \right] + \left[\left(-\frac{81}{4} + \frac{81}{2} \right) - 0 \right]$$

$$total\ area = \frac{81}{2}$$

REPRESENTATIVE RECTANGLE IS HORIZONTAL

Occasionally you encounter a situation in which you have to draw the representative rectangle horizontally rather than vertically.

1 The region, the area of which you are trying to determine, is bounded by the graphs of $x_1 = f(y)$ *and* $x_2 = g(y)$, which intersect at the points with y coordinates $y = a$ and $y = b$.

Note: The height of the representative rectangle is just the right function minus the left function for the shaded region.

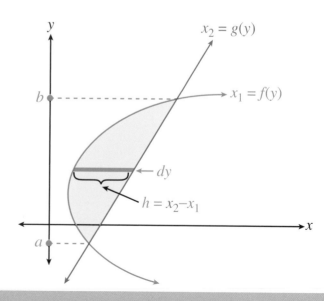

❷ The green representative rectangle has a base of dy and a height of $x_2 - x_1$. Its area is found at right. (Notice that the representative rectangle has a horizontal orientation rather than the usual vertical orientation. If you try to make the rectangle vertical, at the left end of the bounded region, the height would be $x_1 - x_1 = 0$.)

area "rep." rect. $= (x_2 - x_1)dy$

❸ If you add up an infinite number of these very thin rectangles, their sum can be found by the integral at right.

$$Area = \int_a^b (x_2 - x_1)\, dy$$

❹ Substitute $x_1 = f(y)$ and $x_2 = g(y)$ so that the variables within the integrand match the dy term at the end of the integral.

$$Area = \int_a^b \left(g(y) - f(y) \right) dy$$

EXAMPLE 4

Find the area of the region bounded by the graphs of $x = y^2$ and $x = y + 2$.

❶ Find the x coordinates of any points of intersection; these will also serve as your limits of integration.

$$y^2 = y + 2$$
$$y^2 - y - 2 = 0$$
$$(y + 1)(y - 2) = 0$$
$$y = -1,\ y = 2$$

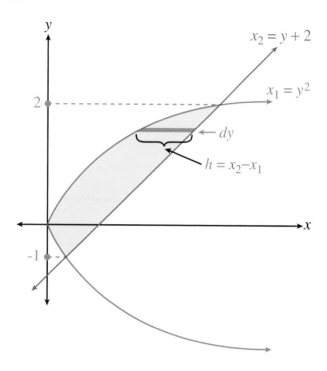

② Sketch the graphs of the two functions, indicating the bounded region the area of which you are going to compute. Also show a representative rectangle for the region, along with its base and appropriate height.

Note: *Notice that the height of the green horizontal rectangle is the* right *function* **minus the** left function.

In the figure:

$x_2 = y + 2$

$x_1 = y^2$

$h = x_2 - x_1$

dy

③ Write the area of the green representative rectangle.

area "rep." rect. $= (x_2 - x_1)\,dy$

④ Set up the integral to find the area of the green shaded region.

$$A = \int_{-1}^{2} (x_2 - x_1)\,dy$$

5 Since you have a dy term at the end of the integral, you need to make the substitutions $x_1 = y^2$ and $x_2 = y + 2$, so that the integrands contain just y terms and numerical values.

$$A = \int_{-1}^{2} \left((y + 2) - y^2 \right) dy$$

$$A = \int_{-1}^{2} \left(y + 2 - y^2 \right) dy$$

6 Evaluate the integral and simplify the resulting computation to find the area of the green shaded region.

$$A = \left[\frac{y^2}{2} + 2y - \frac{y^3}{3} \right]_{-1}^{2}$$

$$A = \left(2 + 4 - \frac{8}{3} \right) - \left(\frac{1}{2} - 2 + \frac{1}{3} \right)$$

$$A = \frac{9}{2}$$

Volume of Solid of Revolution: Disk Method

The region bounded by the graphs of $y = f(x)$, $y = 0$, $x = a$, and $x = b$ is revolved about the x-axis. Find a formula for computing the volume of the resulting solid.

---

❶ At right is a figure showing the bounded region with three thin red rectangles.

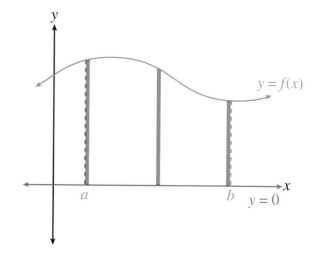

❷ After the bounded region is revolved (or rotated) about the x-axis, it creates a solid as shown at right. Notice that each thin red rectangle traces out a thin red disk (or cylinder).

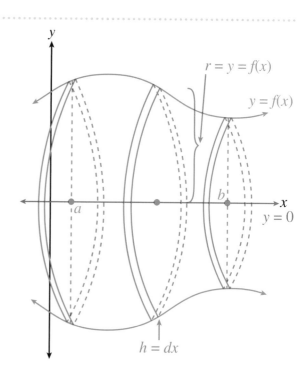

❸ The "representative" disk toward the center of the solid has a height of dx and a radius of $r = y = f(x)$. Using the formula for the volume of a cylinder with radius r and height h, you end up with a formula for the volume of the representative disk.

$$V_{cyl.} = \pi r^2 h$$

$$V_{disk} = \pi y^2\, dx$$

or with all x terms

$$V_{disk} = \pi \left(f(x) \right)^2 dx$$

❹ If you were to find the sum of the volumes of an infinite number of very thin disks, an integral could be used to do that computation.

$$V = \int_a^b \pi \cdot y^2\, dx$$

or

$$V = \int_a^b \pi \cdot \left(f(x) \right)^2 dx$$

❺ When doing a specific problem, it is not necessary to try to sketch the three-dimensional figure. A suggested sketch is shown at right.

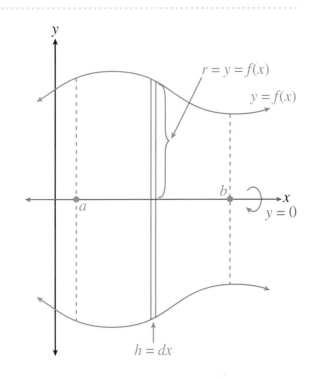

TIP

The *thickness* of the representative disk is really the height of this thin cylinder, in this case $h = dx$. The radius of the disk is always measured from the axis of revolution to the graph being rotated.

6 The appropriate work would be shown as demonstrated to the right.

$$V_{disk} = \pi r^2 h$$
$$= \pi y^2 \, dx$$
$$V_{disk} = \pi y^2 \, dx,$$

then volume of solid is

$$V = \int_a^b \pi \cdot y^2 \, dx,$$

or to get all x terms,

$$V = \int_a^b \pi \left(f(x) \right)^2 dx$$

EXAMPLE 1

The region bounded by the graphs of $y = x^2$, $y = 0$, $x = 1$, and $x = 2$ is revolved about the x-axis. Find the volume of the resulting solid.

1 Sketch a side view of the revolved region. Label the height (looks like the thickness from the side) and the radius of the representative disk.

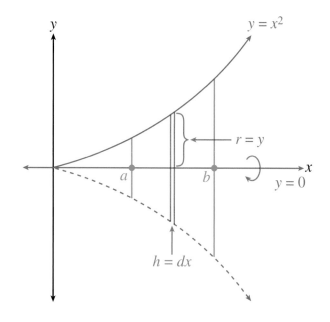

❷ Find the volume of a representative disk.

$$V_{disk} = \pi r^2 h$$
$$= \pi y^2 dx$$

❸ Now you are ready to set up the volume integral.

Notice that the y terms were substituted with appropriate x terms and the constant π was moved out in front of the integral.

$$V = \int_1^2 \pi y^2 \, dx$$
$$V = \int_1^2 \pi \left(x^2 \right)^2 \, dx$$
$$V = \pi \int_1^2 x^4 \, dx$$

❹ Evaluate the integral and plug in the limits of integration to compute the volume of the solid that results.

$$V = \pi \left[\frac{x^5}{5} \right]_1^2 = \pi \left(\frac{32}{5} - \frac{1}{5} \right)$$
$$V = \frac{31\pi}{5}$$

EXAMPLE 2

The region bounded by the graphs of $y = x^2$, $y = 2$, and $x = 0$ is rotated about the y-axis. Find the volume of the solid that results.

❶ Sketch a side view of the figure showing the height and radius of the representative disk.

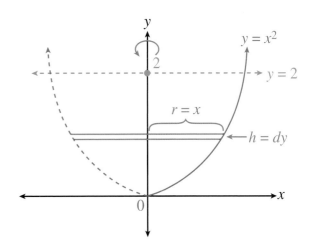

Volume of Solid of Revolution: Disk Method *(continued)*

❷ Find the volume of the representative red disk.

$$V_{disk} = \pi r^2 h$$
$$= \pi x^2 dy$$

❸ Set up the appropriate volume integral, noting that the dy at the end means that the integrand must eventually be in y terms also.

$$V = \int_0^2 \pi x^2 \, dy$$
$$V = \int_0^2 \pi y \, dy$$
$$V = \pi \int_0^2 y \, dy$$

❹ Integrate and substitute the limits to find the volume of the solid.

$$V = \pi \left[\frac{y^2}{2} \right]_0^2 = \pi \left(\frac{4}{2} - \frac{0}{2} \right)$$
$$V = 2\pi$$

EXAMPLE 3

The region bounded by the graphs of $y = \frac{1}{x}$, $y = 0$, and $x = 2$ is revolved about the x-axis. Find the volume of the resulting solid.

❶ Sketch a side view of the solid. Label the radius and height of the representative disk. Note that there is no right hand limit of integration, so in this problem it will be $+\infty$.

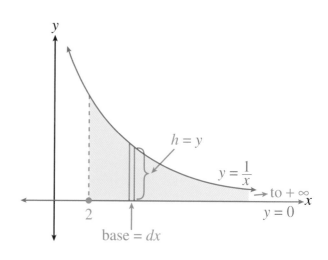

2 Find the volume of the red disk.

$$V_{disk} = \pi r^2 h$$
$$= \pi y^2 dx$$

3 Set up your volume integral and then substitute to get all x terms.

$$V = \int_{2}^{+\infty} \pi y^2\, dx$$

$$V = \int_{2}^{+\infty} \pi \left(\frac{1}{x}\right)^2 dx$$

$$V = \pi \int_{2}^{+\infty} \left(\frac{1}{x}\right)^2 dx$$

$$V = \pi \int_{2}^{+\infty} x^{-2}\, dx$$

4 Integrate and then plug in the limits to get the volume of the solid.

$$V = \pi \left[-\frac{1}{x}\right]_{2}^{+\infty} = \pi \left(-\frac{1}{\infty} - \left(-\frac{1}{2}\right)\right)$$

It's interesting that the area of the region is finite, even though its right hand limit of integration is infinite!

$$V = \frac{\pi}{2}$$

EXAMPLE 4

(axis of revolution not x or y axis)

The region bounded by the graphs of $y = x^2$, $y = 0$, and $x = 1$ is rotated about the line $x = 1$. Find the volume of the solid that results.

❶ Sketch a side view of the solid, noting the height and radius of the horizontal representative red disk. Notice that the left hand limit of integration is just $x = 0$.

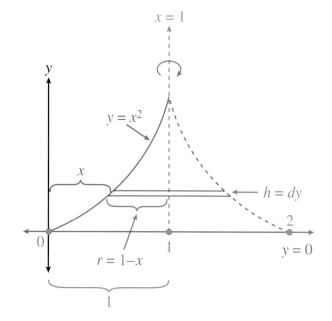

 TIP

Be sure to measure the radius from the axis of revolution back to the original curve.

❷ Find the volume of the representative red disk.

$$V_{dixk} = \pi r^2 h$$
$$= \pi(1-x)^2 dy$$

❸ Set up the integral used to find the volume of the solid.

$$V = \int_0^1 \pi(1-x)^2 \, dy$$
$$V = \pi \int_0^1 (1-x)^2 \, dy$$
$$V = \pi \int_0^1 (1 - 2x + x^2) \, dy$$

4 Since there is a *dy* at the end of the integral, all terms before that must in terms of *y* also. With $y = x^2$, you also have $y^{1/2} = x$.

$$V = \pi \int_0^1 \left(1 - 2y^{1/2} + y\right) dy$$

5 Last, integrate and then plug in the limits.

$$V = \pi \left[y - \frac{4}{3} y^{3/2} + \frac{y^2}{2} \right]_0^1$$

$$= \pi \left[\left(1 - \frac{4}{3} + \frac{1}{2}\right) - 0 \right]$$

$$V = \frac{\pi}{6}$$

Volume of Solid of Revolution: Washer Method

The region bounded by the graphs of $y_1 = f(x)$, $y_2 = g(x)$, the x-axis($y = 0$), $x = a$, and $x = b$ is revolved about the x-axis. Find the volume of the resulting solid.

---

❶ Sketch the bounded region.

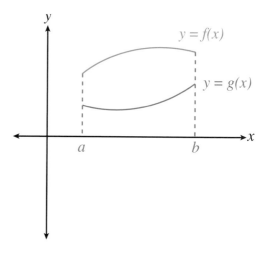

❷ Try to envision the three-dimensional solid with a "hole" through it, which results from rotating the bounded region about the x-axis. It sort of looks like a candle holder on its side. Notice the thin red "washer" with the hole in it.

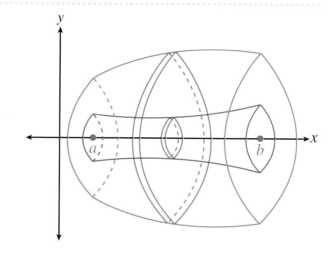

❸ Let's look at just the thin red washer for a moment. It has an inner radius of r and outer radius of R, with a height (or thickness) of h.

Derive a formula for the volume of the thin red washer.

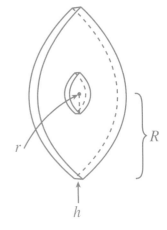

 TIP

R is the radius of the disk and r is the radius of the hole.

$$V_{washer} = V_{disk} - V_{hole}$$
$$= \pi R^2 h - \pi r^2 h$$
$$V_{washer} = \pi (R^2 - r^2)h$$

❹ In the process of doing a problem with a "hole," it is not necessary to try to sketch the three-dimensional version. Just sketch a side view and label the big R, the little r, and the height h.

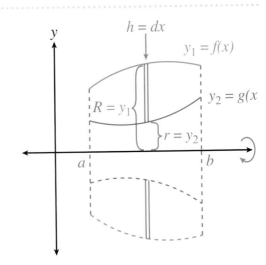

TIP

Both the large radius R and the small radius r must be measured from the axis of revolution; R from the axis to the *outer* curve and r from the axis to the *inner* curve.

5 Substitute the appropriate pieces into the volume of the washer formula.

$$V_{washer} = \pi \left(R^2 - r^2 \right) h$$

$$V_{washer} = \pi \left(\left(y_1 \right)^2 - \left(y_2 \right)^2 \right) dx$$

6 If you were to add the volumes of an infinite number of thin red washers, you could compute that sum by using an integral.

$$V = \int_a^b \pi \left(\left(y_1 \right)^2 - \left(y_2 \right)^2 \right) dx$$

7 With the *dx* term at the end of the integral, you need to make sure that the integrand has only numerical or *x* terms.

$$V = \int_a^b \pi \left(\left(f(x) \right)^2 - \left(g(x) \right)^2 \right) dx$$

EXAMPLE 1

The region bounded by the graphs of $y = x$ *and* $y = x^3$ in the first quadrant is rotated about the *x*-axis. Find the volume of the solid that results.

1 Sketch a side view of the solid; draw in a representative red washer and label its big and small radius as well as its height.

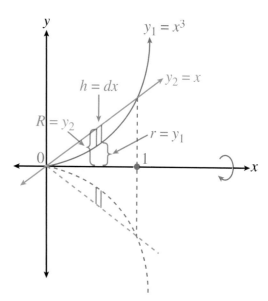

❷ Write the formula for the thin red washer's volume.

$$V_{washer} = \pi \left(R^2 - r^2 \right) h$$

$$V_{washer} = \pi \left(\left(y_2 \right)^2 - \left(y_1 \right)^2 \right) dx$$

❸ Write the appropriate integral to compute the volume of the solid.

$$V = \int_0^1 \pi \left(\left(y_2 \right)^2 - \left(y_1 \right)^2 \right) dx$$

$$V = \pi \int_0^1 \left(\left(y_2 \right)^2 - \left(y_1 \right)^2 \right) dx$$

❹ Replace each y term with its corresponding x term equivalent.

$$V = \pi \int_0^1 \left(\left(x \right)^2 - \left(x^3 \right)^2 \right) dx$$

$$V = \pi \int_0^1 \left(x^2 - x^6 \right) dx$$

❺ Integrate, evaluate, and simplify to get the volume of the desired solid.

$$V = \pi \left[\frac{x^3}{3} - \frac{x^7}{7} \right]_0^1$$

$$= \pi \left[\left(\frac{1}{3} - \frac{1}{7} \right) - 0 \right]$$

$$V = \frac{4\pi}{21}$$

Volume of Solid of Revolution: Washer Method *(continued)*

EXAMPLE 2

The region bounded by the graphs of $y = x^2$, *the x-axis, and* $x = 2$ is revolved about the y-axis. Find the volume of the resulting solid.

❶ Sketch a side view of the solid; draw in a representative red washer, and label its big and small radius as well as its height.

 TIP

The large raduis R is fixed at 2. Only the small radius r is changing.

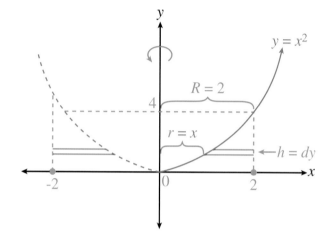

❷ Create a formula for the volume of this particular thin red washer.

$$V_{washer} = \pi(R^2 - r^2)h$$
$$V_{washer} = \pi(2^2 - x^2)dy$$
$$V_{washer} = \pi(4 - x^2)dy$$

❸ Set up the volume integral that enables you to compute the volume of the resulting solid.

$$V = \int_0^4 \pi(4 - x^2)\,dy$$
$$V = \pi\int_0^4 (4 - x^2)\,dy$$

❹ As usual, make sure that the variable within the integrand matches the dy at the end of the integral.

$$V = \pi\int_0^4 (4 - y)\,dy$$
$$V = \pi\int_0^4 (4 - y)\,dy$$

5 Integrate and evaluate the result using the limits of integration.

$$V = \pi[4y - y^2]_0$$
$$= \pi[(16 - 8) - 0]$$
$$V = 8\pi$$

. . . .

EXAMPLE 3

The region bounded by the graphs of $y = x^2$, $y = 0$, *and* $x = 1$ is rotated about the line $x = 3$. Determine the area of the resulting solid.

1 Sketch a side view of the figure; label both large and small radii, as well as the height of the thin red washer.

 TIP

Measure large radius R from the axis of revolution ($x = 3$) to the *outer* curve ($y = x^2$). The small radius r is measured from the axis of revolution ($x = 3$) to the *inner* curve ($x = 1$).

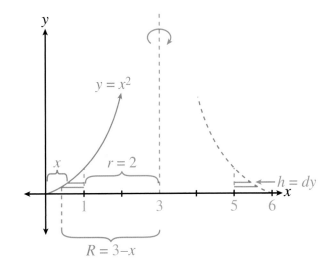

. . . .

2 Write the formula for the thin red washer's volume, substituting for appropriate parts labeled on your figure.

$$V_{washer} = \pi\left(R^2 - r^2\right)h$$
$$V_{washer} = \pi\left((3 - x)^2 - 2^2\right)dy$$

3 Write the integral for calculating the volume of the resulting solid.

$$V = \int_0^1 \pi\left((3-x)^2 - 2^2\right) dy$$

$$V = \pi\int_0^1 \left((3-x)^2 - 2^2\right) dy$$

$$V = \pi\int_0^1 \left(9 - 6x + x^2 - 4\right) dy$$

$$V = \pi\int_0^1 \left(5 - 6x + x^2\right) dy$$

4 Using $y = x^2$ so that $y^{1/2} = x$, substitute so that all variables within the integrand match the *dy* term at the end of the integral.

$$V = \pi\int_0^1 (5 - 6y^{1/2} + y) dy$$

5 Now you're ready to integrate and evaluate using the limits to determine the volume of the solid.

$$V = \pi\left[5y - 4y^{3/2} + \frac{y^2}{2}\right]_0^1$$

$$= \pi\left[\left(5 - 4 + \frac{1}{2}\right) - 0\right]$$

$$V = \frac{3\pi}{2}$$

The region bounded by the graphs of $y = f(x)$, *the x-axis*, $x = a$, and $x = b$ is revolved about the y-axis. Find an expression that represents the volume of the resulting solid.

---

1 To the right is a diagram of the bounded region before being revolved about the y-axis.

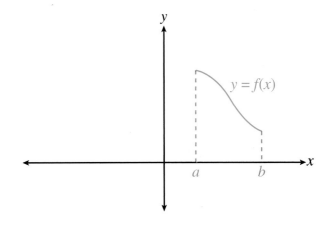

2 After being revolved about the y-axis, the figure shown at right results.

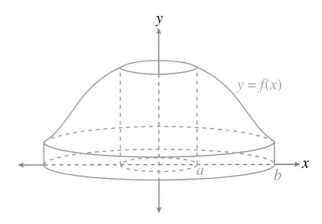

❸ In the new technique, named the "shell method," you find the volume of the resulting solid by finding the sum of an infinite number of very thin "shells" (or pieces of pipe).

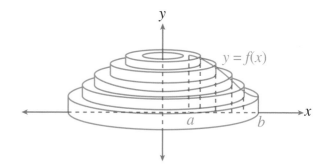

❹ Let's take a look at just one of those many "shells" (or pieces of pipe). The shell has an outer radius, r_2; an inner radius, r_1; and a height of just h. One other dimension, labeled R, is the distance from the axis of revolution to the center of the thin red shell.

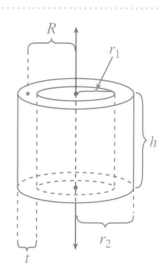

❺ Find a formula for the volume of that thin red shell.

$$V_{shell} = \pi(r_2)^2h - \pi(r_1)^2h$$

This is just the volume of the piece of pipe without the hole, minus the volume of the hole.

6 Play with the formula a bit to rewrite it in another form.

Note: *Big R is really just the average radius. It is measured from the axis of revolution to the center of the shell.*

$$V_{shell} = \pi \left(r_2 \right)^2 h - \pi \left(r_1 \right)^2 h$$

$$V_{shell} = \pi \left(\left(r_2 \right)^2 - \left(r_1 \right)^2 \right) h$$

$$V_{shell} = \pi \left(r_2 + r_1 \right) \left(r_2 - r_1 \right) h \quad \leftarrow \text{Factored the middle term above.}$$

$$V_{shell} = 2\pi \left(\frac{r_2 + r_1}{2} \right) \left(r_2 - r_1 \right) h \quad \leftarrow \text{Mult. and then div. by } 2.$$

$$V_{shell} = 2\pi \left(\frac{r_2 + r_1}{2} \right) h \left(r_2 - r_1 \right) \leftarrow \text{Moved the } h \text{ to the left one term.}$$

$$\downarrow \qquad \downarrow \quad \downarrow$$

$$V_{shell} = 2\pi \quad R \quad h \quad t$$

$$\nearrow \qquad \uparrow \qquad \nwarrow$$

ave. radius   height   shell's thickness

$$V_{shell} = 2\pi R h t$$

7 In the process of doing an actual problem, the figure you sketch will look more like the one at right. R is the radius from the axis of revolution to the center of the shell, and t is the thickness, dx or dy.

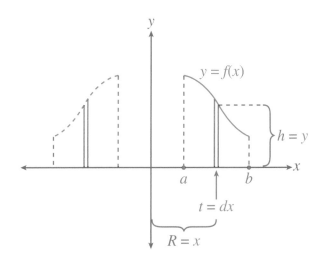

Volume of Solid of Revolution: Shell Method *(continued)*

8 Using the formula $V_{shell} = 2\pi Rht$, substitute the appropriate pieces labeled on your diagram to get a formula for the volume of the thin red shell.

$V_{shell} = 2\pi Rht$
$V_{shell} = 2\pi xy dx$

9 As in other volume techniques, if you were to add the volumes of an infinite number of very thin red shells, the volume of the resulting solid could be determined by using an integral.

$V = \int_a^b 2\pi Rht\, dx$

EXAMPLE 1

The region bounded by the graphs of $y = x^3$, $y = 0$, *and* $x = 1$ is revolved about the y-axis. Find the volume of the resulting solid.

1 Sketch a side view of the solid, with the two thin red rectangles actually representing the side view of a shell. Label the average radius *R*, the height *h*, and the thickness *t*.

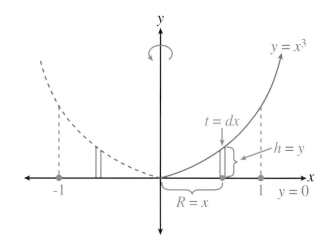

2 Plug the pieces labeled on your diagram into the formula for the volume of a generic shell.

$V_{shell} = 2\pi Rht$
$V_{shell} = 2\pi xy dx$

❸ Set up the integral to find the volume of the solid, substituting to get the integrand in terms of the same variable as the dx at the end.

$$V = \int_0^1 2\pi xy\, dx$$

$$V = 2\pi \int_0^1 xy\, dx$$

$$V = 2\pi \int_0^1 x \cdot x^3\, dx$$

$$V = 2\pi \int_0^1 x^4\, dx$$

❹ Integrate and evaluate the result.

This problem could also have been done using the washer method.

$$V = 2\pi \left[\frac{x^5}{5} \right]_0^1 = 2\pi \left(\frac{1}{5} - 0 \right)$$

$$V = \frac{2\pi}{5}$$

TIP

When using either the disk or the washer method, the "representative" disk or washer is drawn *perpendicular* to the axis of revolution. In the shell method, the "representative" shell is always drawn *parallel* to the axis of revolution. If you have a choice between using the washer or the shell method, it is usually easier to set up the shell method — you need to find only one radius.

EXAMPLE 2

The region bounded by the graphs of $y = x^2$, $y = 0$, *and* $x = 2$ is revolved about the *x*-axis. Find the volume of the resulting solid.

① Sketch a side view of the solid, showing a representative shell.

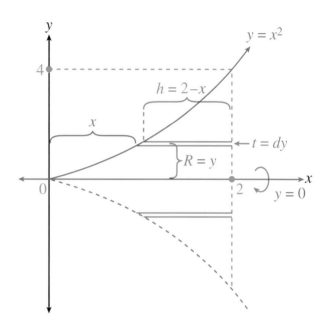

② Plug the pieces labeled on your diagram into the formula for the volume of a generic shell.

$$V_{shell} = 2\pi Rht$$
$$V_{shell} = 2\pi y(2 - x)dy$$

③ Set up the integral to find the volume of the solid, substituting to get the integrand in terms of the same variable as the *dy* at the end.

$$V = \int_0^4 2\pi y(2 - x)\,dy \quad \text{Notice the } y \text{ lim. of int.}$$

$$V = 2\pi \int_0^4 y(2 - x)\,dy$$

$$V = 2\pi \int_0^4 y(2 - y^{1/2})\,dy$$

$$V = 2\pi \int_0^4 (2y - y^{3/2})\,dy$$

④ Integrate and then find the value of the resulting expression.

$$V = 2\pi\left[y^2 - \frac{2}{5}y^{5/2}\right]_0^4$$

$$= 2\pi\left[\left(16 - \frac{2}{5}\cdot 32\right) - 0\right]$$

$$V = \frac{32\pi}{5}$$

This problem could also have been done using the disk method.

EXAMPLE 3

The region bounded by the graphs of $y = x^2$ *and* $y = -x^2 + 2x$ is revolved about the line $x = 3$. Find the volume of the resulting solid.

① Find the x-coordinates of the points of intersection of the graphs of the two equations.

$$x^2 = -x^2 + 2x$$
$$0 = -2x^2 + 2x$$
$$0 = -2x(x - 1)$$
$$x = 0 \quad x = 1$$

② Sketch a side view of the solid and label the appropriate pieces, both the radii and the height.

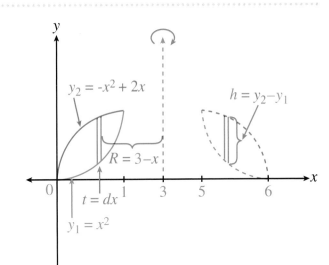

$y_2 = -x^2 + 2x$

$h = y_2 - y_1$

$R = 3 - x$

$t = dx$

$y_1 = x^2$

TIP

The radius R must be measured from the axis of revolution ($x = 3$) to the middle of the shell.

Volume of Solid of Revolution: Shell Method *(continued)*

❸ Find the volume of the thin red shell.

$$V_{shell} = 2\pi Rht$$
$$V_{shell} = 2\pi(3 - x)(y_2 - yi)dx$$

❹ Set up the integral to determine the volume of the solid and then evaluate using the limits of integration.

$$V = \int_0^1 2\pi(3 - x)(y_2 - y_1)\,dx$$
$$V = 2\pi\int_0^1 (3 - x)(y_2 - y_1)\,dx$$
$$V = 2\pi\int_0^1 (3 - x)(-x^2 + 2x - x^2)\,dx$$
$$V = 2\pi\int_0^1 (3 - x)(-2x^2 + 2x)\,dx$$
$$V = 2\pi\int_0^1 (2x^3 - 8x^2 + 6x)\,dx$$

❺ Integrate and then evaluate using the limits of integration.

$$V = 2\pi\left[\frac{x^4}{2} - \frac{8x^3}{3} + 3x^2\right]_0^1$$
$$= 2\pi\left[\left(\frac{1}{2} - \frac{8}{3} + 3\right) - 0\right]$$
$$V = \frac{5\pi}{3}$$

Appendix

In the following, **c** is a constant and **a** is a constant. In cases where **u** appears, **u** is some function of another variable and the **du** is just the derivative of **u**.

GENERAL FORMS

① $\dfrac{d}{dx}(c) = 0$

② $\dfrac{d}{dx}(x) = 1$

③ $\dfrac{d}{dx}[c \cdot f(x)] = c \cdot f'(x)$

④ $\dfrac{d}{dx}[f(x) \pm g(x)] = f'(x) \pm g'(x)$
(Sum/Difference Rule)

⑤ $\dfrac{d}{dx}[f(x) \cdot g(x)] = f'(x) \cdot g(x) + g'(x) \cdot f(x)$
(Product Rule)

⑥ $\dfrac{d}{dx}\left[\dfrac{f(x)}{g(x)}\right] = \dfrac{f'(x) \cdot g(x) - g'(x) \cdot f(x)}{[g(x)]^2}$
(Quotient Rule)

⑦ $\dfrac{d}{dx}[f(g(x))] = f'(g(x)) \cdot g'(x)$
(Chain Rule)

⑧ $\dfrac{d}{dx}(x^n) = n \cdot x^{n-1}$
(Simple Power Rule)

⑨ $\dfrac{d}{dx}(u^n) = n \cdot u^{n-1} \cdot du$
(General Power Rule)

EXPONENTIAL FORMS

⑩ $\dfrac{d}{dx}e^u = e^u \, du \quad \left(\text{in particular, } \dfrac{d}{dx}e^x = e^x\right)$

⑪ $\dfrac{d}{dx}a^u = a^u \cdot \ln a \cdot du$

LOGARITHMIC FORMS

⑫ $\dfrac{d}{dx}(\ln u) = \dfrac{du}{u} \quad \left(\text{in particular, } \dfrac{d}{dx}(\ln x) = \dfrac{1}{x}\right)$

⑬ $\dfrac{d}{dx}(\log_a u) = \dfrac{1}{\ln a} \cdot \dfrac{du}{u}$

TRIGONOMETRIC FORMS

⑭ $\dfrac{d}{dx}(\sin u) = \cos u \, du$

⑮ $\dfrac{d}{dx}(\cos u) = -\sin u \, du$

⑯ $\dfrac{d}{dx}(\tan u) = \sec^2 u \, du$

⑰ $\dfrac{d}{dx}(\csc u) = -\csc u \cot u \, du$

⑱ $\dfrac{d}{dx}(\sec u) = \sec u \tan u \, du$

⑲ $\dfrac{d}{dx}(\cot u) = -\csc^2 u \, du$

INVERSE TRIGONOMETRIC FORMS

20 $\dfrac{d}{dx}(\arcsin u) = \dfrac{du}{\sqrt{1 - u^2}}$

23 $\dfrac{d}{dx}(\text{arc csc } u) = \dfrac{-du}{|u|\sqrt{u^2 - 1}}$

21 $\dfrac{d}{dx}(\arccos u) = \dfrac{-du}{\sqrt{1 - u^2}}$

24 $\dfrac{d}{dx}(\text{arc sec } u) = \dfrac{du}{|u|\sqrt{u^2 - 1}}$

22 $\dfrac{d}{dx}(\arctan u) = \dfrac{du}{1 + u^2}$

25 $\dfrac{d}{dx}(\text{arc cot } u) = \dfrac{-du}{1 + u^2}$

Common Integration Formulas

In the following, **k** is a constant, **c** is a constant. In cases where **u** appears, **u** is some function of another variable and the **du** is just the derivative of **u**.

GENERAL FORMS

1 $\int dx = x + c$

2 $\int k\,dx = kx + c$

3 $\int [f(x) \pm g(x)]\,dx = \int f(x)\,dx \pm \int g(x)\,dx$

4 $\int x^n\,dx = \dfrac{x^{n+1}}{n + 1} + c\,(n \neq -1)$

5 $\int u^n\,du = \dfrac{u^{n+1}}{n + 1} + c\,(n \neq -1)$

LOGARITHMIC FORMS

6 $\int \dfrac{du}{u} = \ln|u| + c$

In particular $\int \dfrac{1}{x}\,dx = \ln|x| + c$

EXPONENTIAL FORMS

7 $\int e^u\,du = e^u + c$

In particular $\int e^x\,dx = e^x + c$

TRIGONOMETRIC FORMS

⑧ $\int \sin u \, du = -\cos u + c$

⑨ $\int \cos u \, du = \sin u + c$

⑩ $\int \tan u \, du = -\ln|\cos u| + c$ or $\ln|\sec u| + c$

⑪ $\int \csc u \, du = \ln|\csc u - \cot u| + c$

⑫ $\int \sec u \, du = \ln|\sec u + \tan u| + c$

⑬ $\int \cot u \, du = \ln|\sin u| + c$ or $-\ln|\csc u| + c$

⑭ $\int \sec^2 u \, du = \tan u + c$

⑮ $\int \csc u \cot u \, du = -\csc u + c$

⑯ $\int \sec u \tan u \, du = \sec u + c$

⑰ $\int \csc^2 u \, du = -\cot u + c$

INVERSE TRIGONOMETRIC FORMS

⑱ $\int \dfrac{du}{\sqrt{a^2 - u^2}} = \arcsin \dfrac{u}{a} + c$

⑲ $\int \dfrac{du}{\sqrt{a^2 + u^2}} = \dfrac{1}{a} \arctan \dfrac{u}{a} + c$

⑳ $\int \dfrac{du}{u\sqrt{u^2 + a^2}} = \dfrac{1}{a} \operatorname{arc\,sec} \dfrac{|u|}{a} + c$

Unit Circle and Some Common Trigonometric Identities

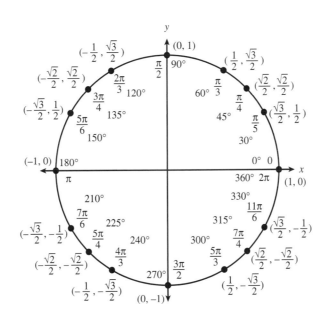

PYTHAGOREAN IDENTITIES

$$\cos^2 x + \sin^2 x = 1 \qquad \tan^2 x + 1 = \sec^2 x \qquad 1 + \cot^2 x = \csc^2 x$$

SUM AND DIFFERENCE IDENTITIES

$$\cos(x \pm y) = \cos x \cos y \mp \sin x \sin y \qquad \sin(x \pm y) = \sin x \cos y \pm \cos x \sin y$$

DOUBLE-ANGLE IDENTITIES

$$\sin 2x = 2\sin x \cos x \qquad \cos 2x = \cos^2 x - \sin^2 x \qquad \cos 2x = 2\cos^2 x - 1 \qquad \cos 2x = 1 - 2\sin^2 x$$

HALF-ANGLE IDENTITIES

$$\cos\left(\frac{x}{2}\right) = \pm\sqrt{\frac{1}{2}(1 + \cos x)} \qquad \sin\left(\frac{x}{2}\right) \pm \sqrt{\frac{1}{2}(1 - \cos x)}$$

Index

A

acceleration function ($a(t)$), 57, 188, 247, 249
acceleration problems. *See* word problems
algebraic functions, 99
algebraic substitution, 237–239
antiderivatives, 195–196
antidifferentiation, 195
approximations, 9
arcsins, 110, 232, 235
area
    bound regions, 205–208
    curves, 9–13, 250–259
asymptotes. *See* horizontal asymptotes; vertical asymptotes
average radius, 277

B

balloon rate of change problem, 72–73, 139–140
bound regions, 205–208, 250–259. *See also* solids of revolution
bus company fare problem, 181–182

C

Chain Rule, 96, 104–109
change of base property, 114
change of variable technique, 216–217
circumscribed rectangles, 9, 11
closed intervals, 31, 155–159
"combo" technique, 235
common definition forms, 60–63
common denominators, 43–45
common logs, 119
complicated natural log expressions, 117
composite continuity property, 30
composite functions, 96, 104–109
composite limit property, 24
concavity, 165–168
conditional functions, 18–19
conical water tank problem, 183–184
Constant Multiple Rule, 79
Constant Rule, 78
continuity, 26–31, 76
cosecant, 100–101
cosine, 97–99, 202
cotangent, 100–101
critical numbers, 146–147
curves
    area, 9–13, 250–259
    tangents, 6–8, 67–69
cylindrical can construction problem, 179–181

D

Δ–E definition of limits of functions, 14–16
decreasing functions. *See* increasing/decreasing functions
definite integrals, 13, 203, 205–208

denominators, 43–45, 47
derivatives. *See also* differentiation
    alternate notations for, 70–71
    analyzing motion on objects with, 57
    analyzing rates of change with, 57, 72–73
    defined, 1
    definition forms, 60–63
    differentiability, 74–76
    of exponential functions, 125–128
    finding equation of lines tangent to curves with, 67
    finding horizontal tangents with, 68–69
    finding points where relative maximums/minimums occur with, 56
    finding slope of tangent lines with, 56, 58–59
    formulas, 66, 97, 114, 119
    of logarithmic functions, 113–122
    Mean Value Theorem, 91–92
    optimizing word problems with, 57
    Rolle's Theorem, 89–90
    rules, 78–88, 93–95, 123–124
    second, 71, 165, 200
    of specific functions at specific numbers, 63–66
    of trigonometric functions, 96–111
differential calculus, 1
differential equations, 240–245
differentiation. *See also* derivatives
    concavity, 165–168
    and continuity, 76
    critical numbers, 146–147
    defined, 55, 60, 65, 80
    extrema, 155–164, 172–175
    finding tangent lines to graphs of functions at points, 143
    horizontal tangents, 144–145
    implicit, 129, 134–141
    increasing/decreasing functions, 148–154
    inflection points, 168–171
    versus integration, 195–196
    logarithmic, 129–133
    overview, 142–175
    rules of, 283–284
    when functions fail to have, 74–75
    word problems, 176–193
direct substitution, 36–37
discontinuity, 28–29
disk method, 260–267
dividing by largest power of variables, 40–42
Double-Angle Identities, 286

E

E, 14–16
e, 113
equations
    differential, 240–245
    graphs of, 250–251
    of lines tangent to curves, 67
    written in implicit form, 134

explicit form, 134
exponential functions
    continuous, 30
    derivatives of, 125–128, 283
    First Fundamental Theorem, 203–204
    integral formulas, 201, 284
    integrals of, 202, 220–222, 284
expressions
    logarithmic, 113–114, 117
    rational, 37, 43–45
extrema, 155–159, 160–164, 172–175. *See also* maximums/minimums
Extreme Value Theorem, 34

F

f, 14
factor and reduce technique, 39–40
first derivative test, 160
First Fundamental Theorem, 203–204
formulas. *See* derivatives; integrals
functions. *See also specific functions by name*
    algebraic, 99
    composite, 96, 104–109
    conditional, 18–19
    power, 201

G

General Power Rule
    and arcsins, 235
    with natural logarithmic functions, 116
    overview, 84–85, 214–219
    with radical trigonometric functions, 101
geometric formulas, 66, 139
graphs of functions
    concavity for, 166–168
    determining limits from, 20–22
    finding maximums/minimums on, 56
    finding tangent lines to at points, 143
    with "holes", 75
    with "jumps", 75
    polynomial, 144
    with sharp turns, 74
    that have two horizontal asymptotes, 53–54
    trigonometric, 145
    with vertical tangent lines, 74
    when intersect once or more, 251–259

H

h (height), 139
Half-Angle Identities, 286
height (*h*), 139
highs, 155. *See also* maximums/minimums
"holes", 27–28, 75, 268–269
horizontal asymptotes
    functions whose graphs have two, 53–54
    functions with *x*-axis as, 49

overview, 48
    of rational functions, 49–52
horizontal rectangles, 256–259
horizontal tangents, 68–69, 144–145

I

implicit differentiation, 129, 134–141
implicit form, 134
increasing/decreasing functions, 148–154
indefinite integrals, 197–200
indeterminate forms, 38–47, 93–95
infinite discontinuity, 29
infinite series, 2–3
infinity, limits at, 48–54
inflection points, 146–147, 168–171
initial conditions, 244
inner radius (*r*), 139, 269
inscribed rectangles, 9–10
integrable, defined, 198
integral calculus, 1
integrals. *See also* integration
    antiderivatives, 195–196
    definite, 205–208
    of exponential functions, 220–222
    First Fundamental Theorem, 203–204
    formulas, 201–202, 226, 232, 284–285
    indefinite, 197–200
    "look-alike", 236
    overview, 1, 194
    Second Fundamental Theorem, 209–210
    that result in inverse trigonometric functions, 232–234
    that result in natural logarithmic functions, 223–225
    of trigonometric functions, 226–231
integrands, 197, 235
integration. *See also* antidifferentiation; integrals
    algebraic substitution, 237–239
    "combo" technique, 235
    definition, 211
    differential equations, 240–245
    versus differentiation, 195–196
    finding area between curves, 250–259
    finding volume of solids of revolution, 260–282
    General Power Rule, 214–219
    limits of, 203, 217–218
    overview, 211–245, 246
    problems, 247–249
    Simple Power Rule, 212–214
Intermediate Value Theorem, 32–33
intervals, 31–33, 148–150, 155–159
inverse trigonometric functions. *See* trigonometric functions

J

"jumps", 27, 29, 75

L

L, 14
ladder sliding down side of building problem, 140–141
large radius (*R*), 272–273
L'Hôpital's Rule, 93–95, 102–103, 123–124, 127–128
light pole and shadow problem, 185–187
limits
    calculating with algebraic methods, 35–54
    calculating with properties of, 23–25
    continuity of functions, 26–31
    determining from graphs of functions, 20–22
    Extreme Value Theorem, 34
    of functions, 4–5, 14–16
    indeterminate forms, 93–95
    of integration, 203, 217–218
    Intermediate Value Theorem, 32–33
    L'Hôpital's Rule, 93–95, 123–124
    one-sided, 17–19
    overview, 1
    Riemann Sums, 9–13
    slopes of lines tangent to curves, 6–8
    of sums of infinite series, 2–3
    terms of infinite series, 2
line tangent. *See* tangents
linear functions, 205–206
log of a power property, 114, 117, 122
log of a product property, 113, 117
log of a quotient property, 114, 117
logarithmic differentiation, 129–133
logarithmic expressions, 113–114, 117
logarithmic functions. *See also* natural logarithmic functions
    continuous, 30
    derivatives of, 119–122
    differentiation, 283
    integration, 284
    L'Hôpital's Rule and, 123–124
"look-alike" integrals, 236
lower approximations, 9
lows, 155. *See also* maximums/minimums

M

maximums/minimums
    on closed intervals, 155
    finding with critical numbers, 146
    on graphs of functions, 56
    relative, 56, 160–164, 172
    word problems, 177
Mean Value Theorem, 91–92
motion on objects, analyzing, 57
multiples, scalar, 30

N

natural logarithmic functions
    derivatives, 113–118, 119
    integrals, 202, 223–225, 284
    products involving, 151–152

negative sign, 141
negative velocity, 191
nth derivatives, 71
numerators, rationalizing, 46

O

one-sided limits, 17–19
open intervals, 31
optimization problems. *See* word problems
outer radius (*R*), 266

P

particle moving along straight line problem, 190–193
polynomial functions
    continuous, 30
    critical numbers of, 146–147
    direct substitution to find limits of, 36
    extrema of, 156–157, 161–162, 172–174
    finding derivatives of, 63–64, 80
    First Fundamental Theorem, 204
    increasing/decreasing functions for, 149–151
    inflection points, 169–170
    integrals, 201
    logarithmic functions of, 121
    one-sided limits for, 19
position function, 188
position problems. *See* word problems
positive velocity, 191
power functions, 201
power limit property, 24
Power Rule, 78
powers, 24, 114–116, 201
problems. *See* word problems
product limit property, 23
Product Rule, 81–83, 101
properties of continuity, 30
properties of limits, 23–25
Pythagorean Identities, 99, 286

Q

quotient continuity property, 30
quotient limit property, 24
Quotient Rule, 86–87, 116
quotients
    continuous, 30
    derivatives of, 99, 116, 121–122
    limited, 24
    of radical functions, 87–88
    of rational expressions, 37

R

r (inner/small radius), 139, 269
R (outer/large radius), 272–273

radical functions
    continuous, 30
    critical numbers of, 147
    derivatives of at specific numbers, 64–65
    derivatives of trigonometric, 101
    direct substitution to find limits involving, 36
    finding derivatives of quotient of, 87–88
    inflection points, 170–171
    limits involving, 42
    logarithmic functions of, 122
    natural logarithmic functions of, 115
radius
    average, 277
    inner/small, 139, 272–273
    measuring, 266
    outer/large, 272–273
    representative disk, 261
rates of change, 57, 72–73, 139–140
rational expressions, 37, 43–45
rational functions
    continuous, 30
    horizontal asymptotes of, 49–52
    indeterminate forms involving, 38
    limits of, 39–41
    one-sided limits for, 18
    relative extrema of functions, 162–164
rationalizing, 46–47
reciprocals, 38
rectangles, 9–11, 250, 256–259
rectilinear motion problem, 190–191
regions. *See* bound regions
related rates problems. *See* word problems
relative extrema. *See* extrema
relative maximums/minimums. *See* maximums/minimums
removable discontinuity, 28
representative disks, 261
representative rectangles, 250, 256–259
reversed differentiation formulas. *See* integrals
revolved bound regions. *See* bound regions
Riemann Sums, 9–13
rocket problem, 188–190
Rolle's Theorem, 89–91
rotations. *See* bound regions

S

• (sum of), 3
scalar multiple continuity property, 30
scalar product limit property, 23
secant, 58, 100–101
second derivative test, 172–175
second derivatives. *See* derivatives
Second Fundamental Theorem, 209–210
sharp turns, 74
shell method, 275–282
Simple Power Rule, 212–214

sine, 97–99, 202
slope, 6–8, 56, 58–59
small radius (*r*), 139, 272–273
solids of revolution, finding volume of
    disk method, 260–267
    shell method, 275–282
    washer method, 268–274
special trigonometric limit property, 24–25
Sum and Difference Identities, 286
sum of (•), 3
sum or difference continuity property, 30
sum or difference limit property, 23
Sum/Difference Rule, 80
sums, 23, 30, 98

T

t. See time
tangents
    derivatives of, 97–99
    to graphs of trigonometric functions, 143
    horizontal, 68–69, 144–145
    slope of, 56, 58–59
    vertical, 74
terms of infinite series, 2
thickness, representative disk, 261
third derivatives, 71
three-dimensional solids. *See* solids of revolution
time (*t*), 137–139, 183–187
trigonometric functions
    continuous, 30
    definite integrals, 206
    derivatives of, 96–111, 283
    evaluated at natural logarithmic functions, 118
    extrema of, 158–159, 174–175
    First Fundamental Theorem, 203
    graphs of, 143
    indeterminate forms involving, 38
    integrals, 202, 226–231, 285
    inverse, 110–111, 232–234, 284–285
    limits of, 22, 37, 40
    products of, 153–154
    special, 25
trigonometric identities, 99, 286

U

unit circles, 97, 285
unknown variables, 136–137
upper approximations, 9
u-substitution technique, 216–217, 237

V

V (volume), 139
variables, 40–42, 136–137, 240–245

velocity, 191. *See also* word problems
velocity functions, 57, 188
vertical asymptotes, 27, 29, 49, 51
vertical tangent lines, 74
volume (*V*), 139
volume of box problem, 177–179
volume of solids of revolution. *See* solids of revolution

W

washer method, 268–274
word problems
    implicit differentiation, 139–141
    optimization, 57, 177–182
    position, velocity, and acceleration, 188–193, 247–249
    rate of change, 72–73
    related rates, 183–187

Want instruction in other topics?

Check out these

All designed for visual learners—just like you!

Read Less–Learn More®
Visual®

Teach Yourself VISUALLY™ Guitar
Charles Kim
0-7645-9642-X

Teach Yourself VISUALLY™ Knitting
Sharon Turner
0-7645-9640-3

Teach Yourself VISUALLY™ Windows XP 2nd Edition
0-7645-7927-4

Look for these and other *Teach Yourself VISUALLY*™ titles wherever books are sold.

Visual®
An Imprint of WILEY
Now you know.